A MODEL-THEORETIC REALIST INTERPRETATION OF SCIENCE

SYNTHESE LIBRARY

STUDIES IN EPISTEMOLOGY,

LOGIC, METHODOLOGY, AND PHILOSOPHY OF SCIENCE

Managing Editor:

JAAKKO HINTIKKA, *Boston University, U.S.A.*

Editors:

DIRK VAN DALEN, *University of Utrecht, The Netherlands*
DONALD DAVIDSON, *University of California, Berkeley, U.S.A.*
THEO A.F. KUIPERS, *University of Groningen, The Netherlands*
PATRICK SUPPES, *Stanford University, California, U.S.A.*
JAN WOLEŃSKI, *Jagiellonian University, Kraków, Poland*

A MODEL-THEORETIC REALIST INTERPRETATION OF SCIENCE

by

EMMA RUTTKAMP

*Philosophical Society of Southern Africa,
South Africa*

KLUWER ACADEMIC PUBLISHERS
DORDRECHT / BOSTON / LONDON

A C.I.P. Catalogue record for this book is available from the Library of Congress.

ISBN 1-4020-0729-9

Published by Kluwer Academic Publishers,
P.O. Box 17, 3300 AA Dordrecht, The Netherlands.

Sold and distributed in North, Central and South America
by Kluwer Academic Publishers,
101 Philip Drive, Norwell, MA 02061, U.S.A.

In all other countries, sold and distributed
by Kluwer Academic Publishers,
P.O. Box 322, 3300 AH Dordrecht, The Netherlands.

Printed on acid-free paper

All Rights Reserved
© 2002 Kluwer Academic Publishers
No part of this work may be reproduced, stored in a retrieval system, or transmitted
in any form or by any means, electronic, mechanical, photocopying, microfilming, recording
or otherwise, without written permission from the Publisher, with the exception
of any material supplied specifically for the purpose of being entered
and executed on a computer system, for exclusive use by the purchaser of the work.

Printed in the Netherlands.

To my father

Egbertus Harmen Bloem

CONTENTS

Preface	ix
Introduction	xi

Chapter One: The notion of "model" in philosophy of science 1

1. The interpretation and use of the notion of "model" in philosophy of science 1
2. Heuristic uses of the notion of "model" 2
3. Statement and non-statement variations of the notion of "model" 5
4. Other methodological uses of the notion of "model": Economics 8

Chapter Two: A model-theoretic account of science 12

1. Introduction 12
2. Terminological note 17
3. The formulation of scientific theories 22
4. The primary semantic interpretation of scientific theories 26
5. The empirical interpretation of scientific theories 29
6. Theories and models in methodological studies of economics 34
7. The problem of "over-determination" of theories by empirical models 44

Chapter Three: The statement account of science 62

1. Introduction 62
2. The two-language view of scientific theories 63
3. Theories and non-observables 65
4. Rules of correspondence 68
5. Ramsey sentences and Putnam's paradox 76

Chapter Four: Variations on the non-statement view of science 91

1. Introduction 91
2. Patrick Suppes's set-theoretic approach to science 91
3. The structuralist programme 97
4. The semantic approaches of Beth, Van Fraassen, and Suppe 108
5. Ronald Giere's naturalistic approach to science 116
6. Nancy Cartwright's "simulacrum" account of science 119

Chapter Five: A model-theoretic realism   141

    1.   Introduction   141
    2.   Reality and science   143
    3.   A modified image of science   145
    4.   The process of science: Paradigms and models   149
    5.   The "abstract" and the "concrete"   158
    6.   Empirical adequacy   163
    7.   Conclusion: The meaning of model-theoretic realism   166

Bibliography   181

Preface

The model-theoretic paradigm of the truth and falsity of sentences under various interpretations of the language emerged in mathematics in the nineteenth century; it carried the day in logic since the first half of the twentieth century; and it gained a foothold in the philosophy of science during the second half of the twentieth century.

In the nineteenth century mathematicians realised that terms they were using in geometry and algebra allowed *multiple interpretations* – and "abstract" mathematics was born. "Straight line" could be interpreted as "large circle on a sphere", and the operation "plus" could be applied not only to various types of number, but also to quaternions, functions, vectors, and matrices. Boole even created a calculus which could be interpreted in terms of propositions, sets, or probabilities.

In spite of nineteenth century mathematics and Boole's work in logic, the early twentieth century saw Russell and the early Wittgenstein inherit from Frege and Peano a mature system of formal logic which was implicitly assumed to be about "the" unique universe of mathematical (especially numerical and set-theoretical) and physical objects. Well-formed sentences were supposed to have meaning to the extent that they report on facts or beliefs about this universe. In this *logical atomist* paradigm – to boldly oversimplify – there was assumed to be just one way to assign denotations to the vocabulary of the language and hence meaning to the sentences. (Belatedly, Saul Kripke's "rigid designators" in *Naming and Necessity* are a relic of this era.)

Logical atomism and the picture theory of meaning did not long survive the rise of *model theory* following Tarski's publications on truth and logical consequence. Truth came to be seen in terms of a relation of satisfaction between a sentence and a particular interpretation of the language. An interpretation is a set-theoretic structure, and the relation of *satisfaction*, determining the assignment of truth values to sentences, is established by freely specifying the denotations of terms and predicate symbols in the domain of the interpretation. Henceforth saying "sentence s is true" can only be elliptic for "sentence s is true under the following interpretation I: ...". This holds also for sentences in natural language, where I is often implicitly provided by pointing, by context, by convention (e.g. the dictionary meaning of a word), etc.

During the second half of the twentieth century some philosophers of science reacted negatively to the new model-theoretic paradigm, some by simply ignoring it, some by considering it irrelevant to the philosophy of empirical science, and some by trivializing it. A prime example of the latter reaction was that by Quine: "Attribution of truth to 'Snow is white' just cancels the quotation marks and says that snow is white. Truth is disquotation." ("Truth" in *Quiddities*.) Note how in "attribution of truth..." the crucial Tarskian, or model-theoretic, "under interpretation I" has gone missing. If it were present (say e.g. I("snow") = cocaine, etc.), then Quine's claim would collapse to nonsense.

But since the 1950's other philosophers of science have taken the model-theoretic paradigm seriously and employed it, for a so-called "non-statement" analysis of empirical theories. Prominent amongst these are the "structuralists" (not to be confused with the homonymous French philosophers) such as Sneed, Stegmüller, and others, and also supporters of the "semantic" view of scientific theories – Beth, Suppes and a number of others. In the course of the last two decades of the twentieth century some philosophers overcame the structuralists' curious denigration of the importance of the linguistic formulation of scientific theories. Now there is a greater appreciation of the rôles of and the interplay between theories as syntactical entities in language, their classes of models as semantical entities in our conceptual and mathematical imagination, and the data empirically extracted from reality to be accommodated within the models.

In this book Emma Ruttkamp demonstrates the power of the full-blown employment of the model-theoretic paradigm in the philosophy of science. Within this paradigm she gives an account of science as process and product. She expounds the "received, statement" and the "non-statement" views of science, and shows how the model-theoretic approach resolves the spurious tension between these views. In this endeavour she also engages the views of a number of contemporary philosophers of science with affinity to model theory. Finally, she defends a sophisticated, referential, model-theoretic *realism* in the philosophy of science as being the appropriate meta-stance most congruent with the model-theoretic view of science as a form of human engagement with the world.

Johannes Heidema
Pretoria, 2001
SOUTH AFRICA

INTRODUCTION

In this text I shall offer a model-theoretic realist interpretation of the processes and products of science built on recent work in the philosophy of science dedicated to analysing the natural sciences in terms of conceptual (mathematical) models of theories and the various semantic relations between such models, theories, and (aspects of) reality. Although analyses of theory-reality and model-reality links have long been a part of philosophy of science, I shall concentrate on the more recent developments concerning these issues in a *model-theoretic* context. Such analyses touch on core questions of the philosophy of science, such as questions regarding the nature of scientific theories, the usual realist inspired questions about the possibility and character of relations between scientific theories and reality, the notion of scientific truth, and, in general, on the nature of scientific progress.

This book has two themes. In one sense I mean it as an introductory text on formal semantic analyses (especially those making use of model theory) in philosophy of science over the past 70 to 80 years. On the other hand I mean this book as an introductory exposition of my account of model-theoretic realism. I join these themes by setting out and explaining a model-theoretic realist account of science against the background of other developments in formal semantics. Mary Morgan and Margaret Morrison (1999) in their book entitled *Models as mediators* make the point (ibid., pp.12,13) that philosophy of science texts in general offers very little information on how models are built. I think there is also very little on exactly what are taken as "models" by different philosophers, and on what the goals and also the consequences of introducing "models" in philosophy of science really are.

A model-theoretic realism offers, via analyses of the structure of scientific theories and the processes of science, a scientific realism that needs, apart from the ontological assumption that the world exists independently of us, very little else from metaphysics. This particular kind of realism is simple although sophisticated, and, as such, implies that although reality is acknowledged to exist "outside" of human practice, this neither means that reality is unknowable nor, at the other extreme of the scale, that science simply mirrors it. By analysing the structure of scientific theories model-theoretically (Chapter 2), it will be shown (Chapter 5) that a model-theoretic approach to science and its processes and products offers the best kind of scientific realism — i.e. a scientific realism with as little metaphysical content as possible.

First a few words on the notion of "science" as it will be used in this text. In principle, my interpretation of science is applicable to all sciences (natural, behavioural, human, social, economic, and so on). I shall in this book however put the emphasis almost exclusively on the natural sciences, and in particular on physics, for the following reason: all of the three main aspects of my model (or interpretation) of science (namely empirical models, interpretative, conceptual models, and theories) are — mostly — more simple and more clearly delineated in the case of the natural sciences (and in physics in particular) than in the other sciences. However, in Section 1.4 I briefly discuss the use of the notion of "model" in economics, and in Section 2.6 I briefly explore the relations between theories and models in methodological studies of economics to illustrate the applicability of a model-theoretic approach outside the natural sciences.

The use of the notion of models is nothing new in either philosophy of science or the (empirical) sciences themselves. Writers such an Achinstein (1968), Hesse (1963), and, more recently, Redhead (1980) have paid much attention to the heuristic uses of models in the development of scientific theories. In his article, entitled *A comparison of the meaning and uses of models in mathematics and the empirical sciences*, Patrick Suppes (1960) reviews the various uses of the notion in mathematical statistics, psychology, economics, and physics. Nancy Cartwright (1989, 1995a, 1995b) also often makes use of the analogies between the ways in which models are used in economics and theoretical physics to illustrate her views of the nature of scientific theories. In Section 1.1 I shall give a brief overview of some of the different interpretations and uses of the notion of "model" in philosophy of science.

In current philosophy of science, many interesting questions centre around the ways in which writers distinguish between and assign roles to theories and the mathematical structures that interpret them and in which they are true, i.e. between scientific theories as linguistic systems and their non-linguistic models. In this context I shall set out my own model-theoretic account of the processes and products of science.

Philosophy of science literature offers us mainly two approaches to the structure of the products of science analysed in terms of linguistic and non-linguistic systems, the "statement" or syntactic approach, and the "non-statement" approach. The statement approach is characteristic of philosophers and logicians like Carnap, Hempel, and Nagel. The advocates of this approach use the tools of mathematical logic to depict theories as axiomatic systems in some well-defined language, and study the syntax and semantics of theories via the proof and model theories of language. The advocates of the non-statement view of science's products generally emphasise the tools of algebra and set theory. This approach originated with Poincaré's work in geometry and mechanics and has started developing through semantic analyses of non-statement reconstructions of certain scientific theories done by Von Neumann (1955), Adams (1959) and Suppes (see McKinsey, Sugar & Suppes (1953), Suppes (1959)), and also Montague (1962).

Defenders of the statement or "received"[1] account of scientific theories depict the rational reconstruction of the language of science as a syntactic system with an axiomatised deductive theory formulated within that system. Its defenders usually characterise theories in terms of two parts. First they identify an abstract formal calculus (a symbolic language) in which the primitive symbols (which in this case are terms that do not have obvious relations with "observation" terms, i.e. so-called "theoretical" terms like "electron", "particle", "mass", and so on) of the theory are set out. The second part of the structure of a scientific theory they depict as a set of rules (called "correspondence rules" by Carnap and "bridge principles" by Hempel) that assigns empirical (observational) content to the logical calculus by providing "co-ordinating definitions" or "empirical interpretations" for at least some of the primitive and defined symbols of the calculus, and in that way — supposedly — establishes partial interpretations for the logical calculus by linking it to the observational language, which is assumed to have a direct interpretation as a description of the "real" world.

Advocates of the "non-statement" approach view the rational reconstruction of the language of science in terms of a syntactic system and a family of interpretations (or models) of that syntax. In opposition to the statement approach's belief that theories are formulated in some (first-order) symbolic language with a set of correspondence rules added somehow, the defenders of the non-statement approach believe in an analysis where the language in which the theory is formulated plays a much smaller role. They hold that foundational problems in the various sciences can in general be better addressed by focussing on the *models* of scientific theories, rather than by reformulating the products of these sciences in some appropriate language, and so they identify a theory with a certain class of mathematical structures.

The non-statement approach has had several branchings since Patrick Suppes has emphasised — against the meta-mathematical musings of the advocates of the received view — the "clarifying" advantages of set-theoretical reconstructions of empirical theories. I shall discuss Suppes's approach in section 4.2. The structuralist programme led by Sneed, Stegmüller, Moulines, and Balzer offers a structural analysis of science, and I shall discuss this programme in section 4.3.[2] The semantic approach offers an examination of the content of theories via Beth's notion of state-spaces and is supported by Van Fraassen, Suppe, and the (naturalistic) view of science offered by Ronald Giere. These notions are discussed in sections 4.4 and 4.5. I shall unfortunately not, due to limitations of space, discuss the "empiricist semantic" accounts offered by, among others, Wójcicki, and Przelecki. These philosophers both concentrate on offering an empiricist semantics for science — Wójcicki was one of the first who applied the notion of intended application in his analysis of the structure of theories — while also working on the problem of "analyticity"[3]. They are joined by philosophes such as Tuomela and Rantala[4], who apply their semantic account of science respectively to problems concerning the nature of theoretical terms and the problems of definability and indefinability in science.

Within the context of my model-theoretic account I also review, in Section 4.6, Nancy Cartwright's views on the nature of "fundamental laws" in science, i.e. the axioms of the deductive set of sentences I refer to as a scientific theory. Cartwright's work is important in my terms since she is currently probably the most influential philosopher of science — writing on the issue of the realism of scientific theories — who does not subscribe to either the statement or the non-statement approaches to science. I do not think that she succeeds either in reconciling or superseding these two approaches though. A model-theoretic account however solves a remarkable number of statement as well as non-statement problems and so supersedes these approaches, reconciling their best features. Cartwright though does pay much attention to general realist issues (think of her pre-occupation with the so-called "lying" laws of physics).

The choice between the statement and non-statement approaches seems trivial as far as theories formulated in first-order languages are concerned. As Theo Kuipers (Kuokkanen, 1994:5) points out, the set of models (that is, structures for which the statements of the theory are true) of theories formulated as a set of statements of some first-order language is exactly the kind of (set of) structures that the defenders of the non-statement approach view as the building blocks of empirical theories. In other words (ibid.), possible interrelations between the two approaches exist in so far as an axiomatised theory may be characterised by a class of interpretations which satisfy it, and an interpretation (or class of interpretations) may be characterised by a set of sentences which it satisfies (and in neither case will the characterisation be unique). Note that in both the statement and non-statement approach the rational reconstruction of a scientific theory is given in terms of an uninterpreted language, which implies, in principle, the possibility of an unlimited number of interpretations of the language. However, in both cases the possible interpretations of the uninterpreted language are limited at least in the following senses. In the statement approach any interpretation must satisfy the axioms of the theory, and in the non-statement approach an interpretation must belong to the described class of structures.

I believe though that clarification of certain core problems in philosophy of science — such as the relations between scientific theories and reality and the notion of scientific truth — is more likely when following the emphasis on models that the non-statement approach offers, while retaining the statement view's analysis of a theory as a (deductively closed) set of sentences. I thus do not go along with either of the two approaches. In the empirical sciences no theoretical entity is ever such a "free creation" of the human mind that its possible links with reality may be discarded in an examination of the various truth relations in which the theory (albeit via its models) stands during the various stages of the scientific process.

I shall set out my own approach in Chapter 2, before continuing to discuss — in Chapter 3 — some of the main points in the statement account of science, and in Chapter 4 I offer explanation and comments on some of the main views in the non-statement school of thought, using model-theoretic realism as a unifying meta-view. In Chapter 5, I shall concentrate on realist issues and offer a model-theoretic interpretation of scientific realism, before concluding on the merits of a model-theoretic realist interpretation of science in the last section of Chapter 5. Since this book is meant as an introductory text in philosophy of science looking at some different applications of formal semantics to realist issues I have tried to minimise the use of formal definitions in logic, although I offer some formalisation of my terminology in Section 2.2.

A brief note on the examples I shall introduce throughout the book. I shall use many well-known examples (such as Newtonian mechanics) to illustrate certain points regarding model-theoretic realism. I choose to do so, since in a comparative sense, different views on the same example seem to be more useful in terms of clarifying points than otherwise. Also I shall refer sometimes (in Chapters 1 and 2) to examples from the philosophy of economics to depict certain aspects of models more clearly. Furthermore I use some very simple examples to illustrate my use of non-monotonic logic (in terms of a minimal model semantics) in model-theoretic realist analyses of the process of science (Chapter 2). For now, though, let us start off by first examining the meaning and use of the notion of "model" in philosophy of science.

On completion of an endeavour such as this, there are always numerous people to thank. I want to thank Johannes Heidema for his comments after proofreading various parts of this text at various times. On a more personal note I thank my son, Franz, for showering me with uncritical love whenever I needed it, and my close family and friends for the various ways in which they helped and supported me towards the completion of this project.

NOTES: INTRODUCTION

[1] Hilary Putnam's term.

[2] Closely related to the structuralist approach is the approach of Ludwig (1990), which I shall not analyse here, but which is worth mentioning, since it is clearly important, although it does not receive much attention in the English literature on philosophy of science.

[3] See Przelecki and Wójcicki (1969).

[4] Pearce and Rantala (1983) claim that they offer a view in which theories are "abstract systems" free of any explicit logical interpretation. This is very interesting, especially since it seems to allow for problems concerning theories too complex to reconstruct in elementary terms, and also since it makes it possible to allow the choice of logic to be an extra-logical (maybe philosophical?) issue.

CHAPTER ONE

THE NOTION OF "MODEL" IN PHILOSOPHY OF SCIENCE

1. THE INTERPRETATION AND USE OF THE NOTION OF "MODEL" IN PHILOSOPHY OF SCIENCE

One intuitive idea of a model is a possible interpretation in which a theory is satisfied in the Tarskian sense, that is, according to Tarski (1956) a model of a sentence in some appropriate language is a possible interpretation of the language in which the sentence of the language are satisfied. This is the basis of model-theoretic analyses of scientific theories, i.e. a model of a theory is a possible interpretation in which all sentences of the theory are satisfied (i.e., in which the sentences are "true"). Model theory was initially developed for explicitly constructed formal languages with the purpose of studying certain mathematical issues, until Evert Beth and others initiated the application of model theory to semantic analyses of so-called "empirical" scientific theories.[1]

Suppes (1960, p.290) points out that the most important distinguishing feature of the relation between a theory and a model of a theory is that a theory is a linguistic entity consisting of a set of sentences, and models are non-linguistic entities in which the theory is satisfied. The notion of "model" is used in widely different ways by philosophers of science and scientists alike. However, if one examines these various interpretations and uses a little more closely, it is possible to find common features in the ways in which the notion of a model is applied that show the advantages of this notion in terms of clarifying the intricacies of the processes and products of science.

Suppes (1960, pp.289,290) remarks that the concept of "model" in the sense of Tarski may be used without distortion and as a fundamental concept in any discipline. In this sense he claims that the meaning of the concept of model in mathematics is the same as in — or very close to that of — the empirical sciences, and finds the difference rather in the various uses of the concept, in the sense that "mathematicians ask a certain kind of question about models and empirical scientists tend to ask another kind of question" (ibid.)[2]. Added to this should be that rather than obscuring the role of models in rational reconstructions of science, these different kinds of "questions" that may be asked about models merely illustrate and strengthen the validity of using the notion of "model" in these kinds of rational reconstruction of scientific theories.

Suppes (in Morgenbesser, 1967, p.57) seems to make much the same point, although perhaps with certain reservations, since he remarks that

> ... quite apart from questions about direct empirical observations, it is pertinent and natural from a logical standpoint to talk about the models of a theory ... [since these] models are highly abstract, non-linguistic entities, often quite remote in their conception from empirical observations.

Therefore, he acknowledges the apparent logical problem underlying questions concerning that which the concept of a model actually has to add to the usual discussions of empirical interpretations of theories. He comments (ibid.) that mostly philosophers find it easier to talk about theories than about models of theories, because mostly, their examples are simple in nature and can be discussed in purely linguistic (first-order logic) terms, while introducing the notion of the model(s) of a theory adds a higher mathematical element to the discussion.

However he (in Morgenbesser, 1967, p.58) points out that when dealing with more complicated theories, like quantum mechanics, and classical thermodynamics, we are not always involved with "simple" examples at all. In these cases he claims (ibid.) we also need the general results of set theory as well as many results concerning the real numbers, since direct formalisation of such theories in first-order logic is rather impractical, given that the degree of complexity of theories of this sort is similar to that of theories studied in pure mathematics. He writes (ibid.):

> In such contexts it is very much simpler to assert things about models of the theory rather than talk directly and explicitly about the sentences of the theory, perhaps the main reason for this being that the notion of a sentence of the theory is not well-defined when the theory is not given in standard formalisation.

Also, some calculations are simply too complex to do in any other context than in that of some model of the theory in question — think for example of a leaf blowing in the wind. To predict where it will come to rest, i.e. to describe its movement, it is possible to use Newton's laws of motion and his law of gravitation[3]. However, the actual calculations would be really complex, and therefore it would be more practical to work in some model of his (i.e. Newton's) "theory" that focuses only on certain of the aspects of the relevant real system.

2. HEURISTIC USES OF THE NOTION OF "MODEL"

Redhead (1980, p.146) comments that early depictions of "theoretical models", such as Achinstein's (1968), almost always implied these models to be false. The reason for this is that mostly, at that stage in philosophy of science[4], models were used in the heuristic sense of suggesting new ways to look at certain problems. Thus the notion of "theoretical model" was generally explained in terms of a set of assumptions about some system, attributing some kind of inner structure to the system, and thus being a "simplified approximation" (Redhead, 1980, p.146) of the system modelled. Also, these models were used — and are of course, still used for these reasons — to justify the production of predictive theories — i.e. theories formulated to explain one group of phenomena (one kind of real system) which then prove to have the ability to predict events concerning the same or another type of system.

In other words, after the positivist reign, most of the initial philosophical reflection concerning models was concerned with the role of models within some kind of "logic of analogy". In these cases it is obvious that models will turn out to be "false" of the system in reality on which the development of the theory in question is focussed, since they are not interpreted in terms of mathematical models (and the empirical reducts) of the theories concerned. Therefore the former usage of the notion of model should not be confused with the latter Tarskian interpretation of the notion in the sense of models "making" the theories they interpret "true".[5]

A related way in which the notion of "model" has also entered into philosophy of science debates is via the well-known use of "iconic" models by Hesse (1963), Achinstein (1968), and Redhead (1980). Both Nagel (1961) and Hesse (1963) view the empirical interpretation of a scientific theory in terms of mathematical and iconic models. Achinstein (1968) defines an iconic model as a set of simplified and approximate assumptions about some system. These assumptions attribute a certain inner structure to the system and are proposed within the context of some more "basic" theory. Hesse (1963, p.19) depicts these models as also possibly exhibiting analogies between the system modelled and some other system. Da Costa and French (1990, p.258) explain these notions in terms of the example of the billiard ball model of kinetic gas theory. Viewed as an iconic model, it offers an image or representation (picture) of a system of gas atoms. There also exists positive analogy in Hesse's terms since there is a similarity between certain aspects of the system itself and certain aspects of the model. The billiard ball model is, however, also a mathematical model of the kinetic theory of gases, in the sense that a semantic interpretation is given to the theory in terms of the billiard ball system such that the axioms of the theory are true under that interpretation.[6]

The main problem with this kind of approach is, as Da Costa and French (1990, p.258) point out, a realist one. The mathematical model is supposed to somehow say something about how the world is (offer an empirical interpretation of the theory), while the same is not expected from an iconic model. Van Fraassen, for instance, ignores these kinds of problem by arguing that scientific theories are indeed nothing more but iconic models of phenomena (observable objects). Da Costa and French (1990, p.259) go on to claim that on the "realist view" it seems that the difference between a theory and models of the theory is that the theory may be true and its models can only be false. This is the same kind of explication that Redhead (1980) offers. The problem is that these kinds of approach view the role of models not semantically, but purely in instrumental terms, that is, in terms of clarifying the axioms of the theory, or extending the domain of a theory, or perhaps even in limiting the domain of theory.

A model-theoretic realist, however, I claim, understands that the role of models in the scientific process is far more important than that. Such a philosopher of science acknowledges that the use of the notion of model in the Tarskian sense as a mathematical interpreting structure is by far the most important one as far as discovering possible links with reality is concerned, and thus such a philosopher only speaks of theories being true in particular models, and not of either theories or models being true or false *per se*, in the literal absolute sense of the word.[7] Do not take me wrong, the powerful role of these kinds of (iconic) models and also of models as "approximations" to theories — Redhead's (1980, p.147) "impoverishments" of theories[8] — in elaborating the domains of scientific theories, cannot be ignored.[9] However, it should be noted that this is a different use of the term than that of the term as a mathematical model, and also that mathematical models can indeed also be used with these kinds of aim in mind.

Redhead (1980, p.162) makes an interesting remark about the role of models in science, that reminds much of Cartwright's (1983) kind of approach. He (1980, p.162) writes

> If we always tried to solve every problem with absolute accuracy, and neglecting no 'accidental' aspects, science would never get started. Science depends on the possibility of ignoring accidents, of isolating certain key features in a situation. These are captured by models, although in the very act of idealisation or approximation we convince ourselves that the model is indeed false. It is moreover in respect of modelling that the imaginative and intuitive element in theoretical physics is most clearly seen ... Modelling is certainly an art, involving a number of logical gaps ....

Of course mathematical models are also idealisations of some real system. The problem that I have with this kind of remark is however that the interpretative models of some theory, despite their potentially idealised nature, may still be found to be related to some real system(s). I argue that it is not "truth" in terms of so-called "false" (or "distorting") idealisations of real systems that philosophers of science should be concerned with at all, but that rather they should concentrate on the Tarskian notion of truth in terms of satisfaction. This kind of (satisfaction) relation establishes (semantic) links between some theory and the models of the theory. It also may determine relations of empirical adequacy in terms of a modified version of Tarskian satisfaction established by (possible) links between the models of the theory and some system(s) in reality. See Chapters 2,3, 4, and 5 for more on this issue.

Redhead (1980, p.147) also discusses the notion of "enriching models". This kind of model is more interesting for my purposes, since it seems to play a role very close to that which I ascribe to the "initial" or "intended conceptual" models in my analysis of the process of science.[10] Redhead claims (ibid.) that these enriching models may come into scientific play whenever a theory seems to be "incompletely specified" such that "considerable latitude" is allowed in the selection of the detailed structure of the theory. He (ibid.) offers the example of axiomatic field theory where any number of fields may possibly satisfy the axioms of the theory, and claims that an "enriching model" may then be introduced to fill in "missing detail". Obviously here the model is not believed to be contradicting the theory, since no "completed" theory has yet been formulated. However, as soon as the model "resembles an exact theory" it is called a "theory" and not a "model" any longer.[11]

Another more heuristically related use of the notion of "model" is the application of the notion in the construction of Gedanken experiments — usually done in the empirical sciences. Here, again, I see this as mainly done in the beginning stage of theory development, except when, as Suppes (1960, p.296) points out, the notion of model is used in arguments against the general plausibility of a theory. In these cases the theory is (conceptually) extended to a new domain (where the scientists expect the results in the new domain to be different from those predicted by the theory) by constructing a model of the theory in that domain. I agree with Suppes that this aspect of the use of models need not however be restricted to Gedanken experiments.[12]

Giere (1991, pp.23-27) also discusses the various kinds of model that scientists use through the course of the scientific process. He identifies four different kinds of model, but I shall here discuss only the two more heuristic ones: Scale models are physical models built to scale and are rarely used, although they may play important roles in the formulation of theories, for example the model of DNA, built by Watson and Crick. Then, secondly, analog models are most useful in the beginning stages of scientific research, since they are usually discarded after theory formulation. He (Giere, 1991, pp.23 24) cites the familiar example of the solar system as an analog model for an atom. In this model the electrons and the nucleus of an atom are said to be analogous to the planets circling the sun. (The idea being that if an atom could be magnified by whatever measure required, one would have an observable object with a structure similar to that of the solar system correspondingly reduced.) This model was very fruitful to scientists working in the first half of the century, especially in the ways in which it showed atoms *not* to be analogous to the solar system, the general point being that "a good analogy leads to its own demise" (Giere, 1991, p.24).

3. STATEMENT AND NON-STATEMENT VARIATIONS OF THE NOTION OF "MODEL"

Wójcicki (1979, p.158) sums up the different usages of the notion of model that philosophers of science should keep in mind as follows:

> ... while the semantic concept of a model is in common use among logicians and mathematicians, empirical scientists almost unanimously opt for the 'mathematical' meanings of 'model' in its *syntactic* (a set of equations) or *semantic* (a mathematical entity which is to represent an intuitive concept) sense.

The interpretation of the notion of a model in the Tarskian sense does not really play a big role in statement depictions of the nature of scientific theories or the process of science. For instance, as an example of a model, Nagel (1961, p.94) offers Bohr's theory of the atom and comments that "... the *theoretical* notion of an electron jump is linked to the *experimental* notion of a spectral line". Nagel claims that more than merely being an example of correspondence rules, the above example also illustrates what he means by a model or interpretation of a theory. He (ibid., p.95) writes:

> The Bohr theory is usually not presented as an abstract set of postulates, augmented by an appropriate number of rules of correspondence for the uninterpreted nonlogical terms implicitly defined by the postulates. It is customarily expounded, as in the above sketch, by way of relatively familiar notions, so that instead of being statement-forms the postulates of the theory appear to be statements, at least part of whose content can be usually imagined. ... in such an exposition the postulates of the theory are embedded in a model or interpretation. It should nevertheless be clear, despite the use of a model for stating a theory, that the fundamental assumptions of the theory provide only implicit definitions for the theoretical notions employed in them. ... Moreover, the presentation of a theory by way of a model does not make any less imperative the need for rules of correspondence for linking the theory to experimental concepts. Although models for theories have important functions in scientific imagery ... models are not substitutes for rules of correspondence.

He claims (ibid., pp.96,97) that all expressions that form part of the formulation of models can be called "meaningful", and so a theory supported, as it were, by a model can be said to be "completely interpreted", since it is then implied that every sentence of the theory is a meaningful statement. But then Nagel also (ibid.) claims that models may mislead us as far as "the actual content of a theory" is concerned, since a theory may receive (ibid.)

> ... alternative interpretations by way of different models; and the models may differ not only in the subject matter from which they are drawn but also in important structural properties.

Depicting a theory with the help of a model, he (ibid.) warns — despite his claim above that a theory "supported by a model" is "completely interpreted" — does not imply necessarily that "... the theory is thereby automatically linked to experimental concepts and observational procedures". He says that this would depend on the structure of the model, but does not elaborate. Indeed, as we shall see in Chapter 2, in a model-theoretic context establishing links between models and systems in reality does depend on the kind of model we are dealing with at the time. However, within the latter context, models are viewed as offering an alternative to typical statement view sets of c-rules. More about this in Chapters 2 and 3.

The followers of the broad non-statement approach (Suppes, Van Fraassen, Suppe, Giere, Wójcicki, Przelecki, and the various structuralists) have among them some variation on the use of the notion of model. In general all of these support the idea that a scientific theory has to be logically reconstructed in terms of a description of its set of models, which is taken as the structures in which the theory's domain can be modelled. They also view models in the Tarskian sense in so far as these models are taken to be relational structures "for which all the sentences of a theory express true properties about the structure when the latter acts as an interpretation of the theory" (Da Costa & French, 1990, p.250).

Przelecki (1969) analyses the structure of scientific theories in terms of theories formulated in first-order predicate logic. He wants to develop an empirical semantics for such theories, and thus needs to offer some analysis of the empirical interpretation of the basic predicates of scientific theories. He offers this analysis in model-theoretic terms:

> In part, an interpretation of a given language is identified with a model theoretic entity — a model M of language L. M assigns to each non-logical constant of L a suitable set-theoretic entity as its denotation. Thus, e.g., a one-place predicate of L is interpreted by M as denoting a certain set of objects from the universe of L. ... my ultimate aim [in (1969)] has been to answer the question ... how is an empirical interpretation possible ... (Przelecki, 1974:401, 402).[13]

I shall here, in this section, only comment further on Van Fraassen and Giere's uses of the notion of "model", since the intricacies concerning the structuralist interpretation of the notion are best discussed as part of their programme — which shall be done in Section 4.3 — and Wójcicki's (1979) notion of theoretical models differs only marginally from what follows.

Van Fraassen (1980, p.43) states clearly that any structure that satisfies the axioms of a theory by making the axioms true is called a model of the theory. He (Van Fraassen, 1980, p.44) points out that scientists usually use "model" as referring to "type of structure" rather than a specific structure, in the sense that some parameters are left unspecified in their description of the structure — he gives the example of the Bohr model of the atom which was intended to fit hydrogen atoms, helium atoms, and so on. Van Fraassen's own use of the term is directed by his notion of empirical adequacy and thus focuses on the identification of empirical substructures of models of theories, since establishing the empirical adequacy of a theory depends on being able to show that the structures which are described in experimental and measurement reports concerning the theory[14] are isomorphic to the empirical substructures of a model of that theory. A more in-depth discussion of Van Fraassen's semantic approach follows in Section 4.4, but I want to point out briefly that my notion of empirical models — see Chapter 2 — is very close to his notion of empirical substructures, although I still want to be a realist even if I acknowledge the role of empiricism in science. More about this in the following chapters.

I have referred in the previous section to Giere's identification of the various uses of the notion of "model". The semantic notion of "model" that he discusses is in terms of what he refers to as "theoretical models". According to him (Giere, 1991, p.24), this interpretation of the notion of "model" is the kind most often used in science. Giere (ibid.) uses the analog of maps to set out the characteristic features of theoretical models. A map is not the same thing as the thing it represents, although some kind of relationship does exist between a map and the object "mapped". This relationship is according to Giere (1991, pp.25,26) a relationship of similarity. A theoretical model (map) exhibits a certain similarity of structure with the thing mapped. Although theoretical models are incomplete in the sense that they exhibit only selected features of their subject (as maps do), still they are "... similar in some specifiable respects and to some specifiable degree of accuracy" (ibid., p.26). I shall comment more fully on the notion of "theoretical model" in my discussion of Giere's semantic approach in Section 4.5, and thus suffice it to remark here that although Giere seems to see some kind of semantic or perhaps interpretative link between theoretical models and certain aspects of reality, he still claims these models to be "false", and thus creates a few (at least in a model-theoretic context) uncertainties concerning his notion of "similarity of structure" referred to above.

A last comment here: Suppes (1960, p.291) claims that the set-theoretical use of the notion of "model" is more primary than the empirical scientific use of it. In a model-theoretic context the so-called "physical model" used in the empirical sciences[15] may perhaps play a role at the lowest most concrete level of the construction of what I refer to as the "conceptual intended model" in the beginning stages of theory formulation (see Chapter 2). However, the Tarskian notion of "model" — equivalent to Suppes's "set-theoretical" notion — is much more important and is applicable both to the role I assign to models in the initial stages of theory formulation and the role I assign them in the interpretative or application stage of the theory's development.[16]

4. OTHER METHODOLOGICAL USES OF THE NOTION OF "MODEL": ECONOMICS

Hausman (1992, p.71) states that the "logician's notion of a 'model' is *not* what economists mean when they talk of models". Rappaport (1998) gives as examples of models as "simplified representations", the Tiebout model, the supply and demand model, the Walrasian or abstract general equilibrium model, and the Keynesian income/expenditure model. (I suspect that some of these models may, model-theoretically speaking, be of different types, but more about that later.) Apart from the distinction between models used as analogies and mathematical models, what often happens in the applied sciences is that the notions of "model of a theory" and "theory of the model" are not always distinguishable. Suppes (1960, p.289) points to an example of such confusion often found in the behavioural sciences and mathematical statistics. Here the word "model" is usually used to mean the set of quantitative assumptions of the theory, which in logical terms are synonymous with the axioms of the theory. The model is here a *linguistic* entity in contrast to the *Tarskian* use according to which a model is a *non-linguistic* entity in which a theory is satisfied.

It seems also if modern economics and econometrics are taken into account, that economists are in general likely to confuse the notions of "theory" and "model" such that their theories are mathematical models in the Tarskian sense, and their models are the linguistic expressions that the supporters of the statement approach call theories. This is supported by Harrod (in Wolfe, 1968, p.189), who concludes that he would prefer that in science (including social science) "... the word 'model' should be confined to formulae relating to posited entities, viz. to entities that we can never directly observe and about the very existence of which we cannot be sure". And he (ibid., p.190) goes on to say that

> If we want to bring 'models' into economics, ... to keep them meaningful, we might confine the term to a system of equations, not all of which are tautologies. Some at least might have adjustable parameters. It might be made a condition for the use of the word that some equations explicitly omit to take account of fringe influences.

Hausman's (1992, p.75) comments on the nature of economic models also broadly confirm this suspicion. He writes:

> Although some economic models are also models in other senses of the term, I know of none in theoretical economics which cannot be characterised as a [set-theoretic] predicate or as a definition of a predicate. Taking models as definitions permits one to develop a cogent interpretation of economic models. Note that this sense of 'model' is distinct from the logical positivist's notion. In their notion, a model is an interpretation of the sentences of a theory such that they all come out to be true. Models of the sort I am talking about, in contrast, are definitions and are constituted by sets of assumptions. They have nothing to do with the semantic interpretation of theories as sets of sentences.[17]

Note however that as early as 1968, Harrod (in Wolfe, 1968, p.180) points out the difference he sees between physics and economics and the various uses of "model" in the two sciences as resting on the fact that in physics, because there is no direct access to reality, "one can *never* compare the model with the reality", while in the case of economics, "all the entities with which economics deal are, in principle at least, directly observable" (ibid.,p.183). This I find difficult to agree with. Hausman's (1992, chapter 4) approach seems to bear me out here — he (ibid., p.81) remarks for instance that

> Even though models in economics need not be as abstract as those which characterise mainstream theorising, they will never apply to economic reality cleanly. Insofar as one has hopes for economic theory at all, there will always be some need to divorce conceptual development and empirical application. 'Unrealistic' model-making is unavoidable for theoretically inclined economists.

Well, Cartwright's (1999) point is that neither is this avoidable in physics, but in neither case is this cause for distress among realists, as I argue in the rest of this text, especially in Chapters 2 and 5.

In Rappaport's (1998, p.128) terms an economic model is a mini-theory. He (ibid.) claims that the construction of a model implies the making of a set of assumptions and the logical (or mathematical) derivation of additional statements, such that all these derived statements form some deductive system, which, then simply is the model. (This is almost a textbook definition of a syntactic view of a scientific theory.)

A last few remarks by Roy Weintraub (1974) on the theory-model identity crisis. He (ibid., p.18) writes:

> The [general equilibrium] theory ... [does] not appear to have much flesh on it. It ... [does] not have much to say about how a poor man can survive in a rich country; it will not be able to answer a critic of a capitalists's society's spending on military projects. Rather a theory ... provide[s] the bare outlines of an ascetic's view of the economic process. ... If the economist is to be more than an observer or a data-collector [s]he must explain economic phenomena. To do this requires models, simplified of necessity, of the processes [s]he wishes to study.

Hausman (1992, p.277) holds that theories — or what is actually Giere's theoretical hypotheses — make testable claims about the world. These claims are simply that specific theories are, indeed, true of the world, while models — Giere's theoretical models — are either 'trivially' (i.e. analytically) true or neither true nor false of the world. Models (ibid.), in these terms, have to be assessed mathematically and conceptually and their point is the conceptual exploration of theories.

Morgan (in De Marchi, 1988, p.318) reviews the modelling process in economics à la Nancy Cartwright as follows: economists start off with as accurate as possible accounts of the behaviour of some system of phenomena in reality that is "prepared" into a (descriptive) model which "matches" a mathematical representation "coming" from the theory. In this way the behaviour of phenomena (via Cartwrightian phenomenological laws) come under the governance of the fundamental laws of the theory. As such, as in Cartwright's and my own terms too, models become the "critical intermediaries" between theories and reality (real systems): Morgan (ibid.) also states that models are at the same time models of theories and models of phenomena

> ... because the model is where two elements are fitted together, but they are also relevant for intervention [in terms of experimentation and empirical tests], because it is models (not theories) which are used as instruments for putting ... [science] at work in the world.

In this way we have the "fitting together of theory elements and data elements into a representation that mixes fiction and fact" (ibid.)[18]

Chiang (1974, p.8) refers to an economic model as a "theoretical framework" of "primary factors and relationships relevant" to a specific problem.[19] For Chiang (ibid., p.35) typically the construction of a model implies the following steps: selection of appropriate endogenous and exogenous variables for inclusion in model; reconstruction of the set of theoretical assumptions regarding the human, institutional, technological, legal, and behavioural aspects of the environment affecting the working of these variables into mathematical equations; and then, an attempt to derive a set of logical consequences from these equations by the application of some mathematical operation.

Whether in economics or the natural sciences though, it is obvious that all models have one common feature: they are "mapping" the elements of the (real) system modelled onto the model.[20] Let us now turn to the issue at hand — namely to work out a model-theoretic *realist* interpretation of science.

NOTES: CHAPTER 1

[1] For a good introduction into model theory for first-order languages, read *Model theory* (1990, 3rd. edition) by C.C. Chang and H.J. Keisler.

[2] Suppes (1960, 290, 291) writes: "To define formally a model as a set-theoretical entity which is a certain kind of ordered tuple consisting of a set of objects and relations and operations on these objects is not to rule out the physical model of the kind which is appealing to physicists, for the physical model may be simply taken to define the set of objects in the set-theoretical model".

[3] I shall from now on sometimes speak of Newton's "theory" when referring to his laws of motion and his law of gravitation.

[4] See also Hesse, M. 1963. *Models and analogies in science*. As Redhead (1980, 149) remarks, Hesse does however point to the role of mathematical models in the development of theories, although she pays much more attention to the role of models in science in terms of analogies.

[5] I am not implying here of course that these (Tarskian) models are "necessarily true" depictions of systems in reality, but simply wish to point out that here at least the possibility of such a turn of events is possible, albeit sometimes by rather complex means. I shall come back often to this point.

[6] Another example that illustrates the relation between Hesse's analogous models and so-called iconic models is given in Da Costa & French (1990, p.250). They (ibid.) write: "In a nucleus ... there are too few particles for a statistical treatment, and there is no overriding centre of force which would enable us to treat the forces between nucleus as small perturbations. For this reason, physicists have fallen back on the 'as if' methods of attack, also known ... as the method of nuclear models. This method consists of looking around for a physical system, the 'model', with which we are familiar and which in some of its properties resembles the nucleus. The physics of the models are then investigated and it is hoped that any properties discovered will also be properties of the nucleus. ... In this way the nucleus has been treated 'as if' it were a gas, a liquid drop, an atom, and several other things".

[7] The arguments that Da Costa and French (1990, p.260) offer in support of their claim that models can only be false do not have anything to do with mathematical models, but only with the use of models as iconic models. Therefore they can only be allowed to conclude that *iconic* models are false, which after all is rather obvious, given the analogous "as if" role of these models.

[8] Redhead (1980, p.147) remarks that models are used as "impoverished theories" if a theory is so complicated that it is very difficult to draw any kind of empirical conclusion from it, since comparisons between the theory and experimental results prove to be too complex. He also shows clearly that the role of these kind of models is not to be confused with the role a Tarskian model plays in the process of science: — "... the important ingredient

... [is] that [the model] and [the theory] logically contradict each other, so that we believe [the model] to be false insofar we believe [the theory] to be true" (Redhead, 1980, p.147).

[9] Note that also theories which are proved somehow empirically inadequate through experimental or some other type of empirical investigation, turn into "impoverishments of theories" — for example (Redhead, 1980, p.147) Maxwell's kinetic theory of gases is now known as the billiard ball model of gases.

[10] See Ruttkamp (1997a), as well as Chapter 2 of this text.

[11] In my scheme of things, only the "intended" model of some theory has the potential to develop into a "full-fledged theory". See my explanation and discussion of these notions in Chapter 2.

[12] See Suppes (1960, p.296) for his reference to Mach in this context.

[13] Tuomela (1972b, 1974) goes to great lengths to point out the problems involved in Przelecki's assumptions concerning the fixing of the universe of the language L in advance — which Przelecki claims is necessary to do in order to "explain the fact of empirical interpretations" (Przelecki, 1974, p.404).

[14] See Van Fraassen (1980, pp.45ff.), and Chapter 5 of this text.

[15] Suppes and Giere and Wójcicki all seem to think that scale models and — even more physical perhaps — models of aeroplanes and cars are at least part of the notion of a "physical" model. In my terms part of the conceptualising that culminates in the intended model may well be directed towards such a type of model — or not, depending on the particular line of research in question. (See Chapter 2.)

[16] Note that "physical" here does not necessarily mean concretely physical, but merely serves to show the more "direct" link with the real system of reality being examined. Any activity ending in the construction of a model is conceptual in the sense that various activities of abstraction and even idealisation are performed. As far as the very few times that an actual concrete model is built go — think for example of the model that Watson and Crick (see Giere, 1991) built of the DNA molecular structure — I would say that, usually, even then, at the same time some kind of (interpretative)conceptual model is also created.

[17] See Hausman (1992, p.77) for a table of differences between models and theories.

[18] See also Morgan (1990) and Hoover (1994).

[19] Chiang (ibid.) also points out that economic models do not necessarily have to be mathematical, but if they are, they consist of a set of mathematical equations describing the structure of the model, that gives mathematical form to the set of analytical assumptions adopted by the model, via the relation of variables to each other in certain specific ways. Then, via some mathematical operation to these equations, a definition of a set of conclusions which logically follows from the assumptions of the model, can be formulated.

[20] "Model" in *The Fonatana dictionary of modern thought* (1977).

CHAPTER TWO

A MODEL-THEORETIC ACCOUNT OF SCIENCE[1]

1. INTRODUCTION

What is it that philosophers can or should say about science? Should we explain the actions of scientists? Should we explain the methodology of science? Should we explain the process and progress of science? Should we study the knowledge claims offered by scientists and decide which of these are so-called "scientific" knowledge claims? Should we study the history of science and its influence on current scientific practice? What — if anything — should we say about reality and the links — again, if any — that exist between science and reality?

In his article on philosophy of science in *The Oxford Companion to Philosophy* (1995) David Papineau describes philosophy of science as being divided into what he calls the "epistemology of science" and the "metaphysics of science". He claims the former to be concerned with questions about the justification and objective nature of scientific knowledge and the latter to be occupied with "philosophically puzzling aspects of the reality uncovered by science" (ibid.:809). I agree that philosophy of science consists of, among other things, an epistemology of science. I would also agree to describe such an epistemology in terms of justification for scientific knowledge, under condition that it is understood that a scientific epistemology should also — perhaps before anything else — offer an analysis of the nature of scientific knowledge, and the process by which we come to such knowledge claims.

As far as a metaphysics of science is concerned though, rather than any kind of metaphysical musings concerning philosophical questions about the "philosophically puzzling" parts of reality that science "uncovers", a philosophy of science should offer an ontology of *science*. If we do assume — as Papineau obviously does — that science can "uncover" certain aspects of reality a philosophy of *science* should surely focus rather on the structure of such an enterprise than that of reality? We have long moved beyond the time of so-called "natural philosophy", and so should acknowledge that philosophy of science (as it should) is first "about" science and not first "about" reality. Of course, if it is assumed that science "uncovers" aspects of reality, we have to say something about reality as it is linked to science, but never should we, as philosophers of science, try to write an ontology or (ontologies) of reality.

But, what then of realism? I shall argue in this text that the only kind of realism possible is indeed one that focuses more on science than it does on reality. For a workable scientific realism the only necessary and sufficient condition that is concerned with the nature of reality is, simply, the assumption that there exists such a reality independent of whatever happens in science. All the other conditions for such a realism should be concerned with science — its nature, its progress, and its justification. For, how can we claim that science is about reality if we do not study science and its processes? Perhaps it seems as if this question implies an equally strong motive for studying reality. However the assumption of the independent existence of reality takes care of that part of the question, simply because, in the end, a philosophy of science is about science, and science is about reality. When it comes to studying the links between science and reality, the original existential assumption concerning reality may be augmented by certain general ("common-sensical")[2] ontological orientations concerning the nature of reality; for instance its complex and rich nature, as well as its similarities and regularities, should be taken into account. But the bottom line remains the same since, even then, the reason for these ontological orientations is that the way science links itself to reality cannot be motivated or understood without these orientations.

The underlying reason for the complexity of the science-reality connection is that science and reality represent different logical classes and therefore they cannot simply be compared to each other directly — especially not in a one-to-one relation of correspondence — without committing a chain of serious category mistakes. A discussion of the process of science in model-theoretic terms, that analyses the development of scientific theories from their origin to their applications forms the foundation of a model-theoretic realism. In this chapter, in what follows, I shall analyse science and its enterprises model-theoretically. In Chapter 3 I shall examine the statement account of science in the context of model-theoretic realism and then, in Chapter 4, I shall move on to discuss some of the main non-statement analyses of science, and also comment on and discuss Cartwright's analysis of science and her specific kind of realist oriented problems. Finally I shall discuss the realist features of science in model-theoretic terms (Chapter 5).

I am advocating a philosophical approach to science which introduces a mediating factor (the role of "models") between scientific statements and "real" objects and events. I shall show that such an account of science examines and make sense of the fact that there is no unique kind of empirical linkage between theories and real systems, but rather a variety of many-to-many links between theories, their (interpretative and empirical) models, and systems in reality. A model-theoretic account of science implies an articulation, a kind of "doubling up" — by splitting up — of the relations between scientific statements and objects in reality, because it implies these relations have to be regarded as "filtering" through the models in the case of both the formulation of a scientific theory and its application. In this view a conciliation of the statement and non-statement views of science is achieved.

More specifically, the model-theoretic analysis of the process of science that I propose is done in terms of an account of science that offers a rational (conceptual) reconstruction of the "life" of a typical scientific enterprise, and is offered against a stratified view of the process of science. This stratification is three-fold: it consists of an empirical level, a middle conceptual (interpretative) level, and a linguistic level.[3] The terms of the first level are very particular, those of the second are more general, although still specific, while the terms at the final level are at the highest level of generality. The distinctions and relationships between the levels are complex (and often interchangeable), and so the process of science will be presented as a development in time from the first level through the middle level to the third level, and then back again.

The "purely empirical" level I interpret as based on the existence of various systems in reality and consists of our interactions with them, while I view the final level as a level of linguistic systems at which a scientific theory is formally formulated and suitably expressed in some appropriate language. The middle conceptual level is a very complex one in the sense that it has various facets that, in their turn, may be seen in terms of a certain kind of hierarchy. It is at this level that models both interpreting scientific theories and making them true are constructed, and, also at this level, the issue of adequate reference to real systems is touched on for the first time. I claim the latter relations of adequacy to be the result of various scientific experimental and observational activities, which may lead to the establishment of a relation of isomorphism between a special kind of substructure of some model of a theory and some "empirical" model of some real system (conceptualising the experimental relations executed in that system).

Before I get on with a discussion of my model-theoretic view, some comments on the Kuhnian meaning of the notions of "paradigm" and "model". In the Postscript to *The structure of scientific revolutions* (1970), Kuhn admits that at least two main and "very different usages" of the term paradigm are possible. I shall in what follows briefly discuss these two usages — paradigms as "disciplinary matrices" and as "exemplars" — in terms of a model-theoretic account of science. I agree that the *Structure*'s notion of paradigm in the sense of the context within which theories are interpreted and applied is a valid and unmistakable part of the conceptual processes of science. As shall be explained in the remaining sections of this chapter, my usage of the notion of model identifies models as components of these kinds of paradigmatic context.

Kuhn (1970, p.175) claims that the two main senses in which he meant the notion of "paradigm" to be interpreted are:

- as standing for the "entire constellation of beliefs, values, techniques, and so on shared by the members of a given community";

- and as denoting "one sort of element in that constellation, the concrete puzzle-solutions which, employed as models or examples, can *replace explicit rules* as a basis for the solution of the remaining puzzles of normal science".

Kuhn (ibid., p.182) acknowledges that scientists would rather speak of sharing a theory, or a set of theories, than of sharing a paradigm or a set of paradigms. This led him to formulate the term "disciplinary matrix" to denote that which is shared by members of a scientific community and that accounts for the success of their communication and their agreement about professional judgements. He (ibid.) explains that "disciplinary" refers to the common possession of members of a specific discipline, and that "matrix" refers to the fact that what is shared is an ordered system which consists of elements of various kinds, each needing further specification. Disciplinary matrices (ibid., p.184) consist of mainly three components:

- first, symbolic generalisations — either formalised, or expressed in words, like "elements combine in constant proportion by weight";

- second, what he referred to as the "metaphysical" parts of paradigms and which he explained as commitments to beliefs such as "all perceptible phenomena are due to the interaction of qualitatively neutral atoms in the void, or, alternatively to matter and force, or to fields" and which he described as beliefs in models; and,

- lastly, values such as predictive power, simplicity, consistency, plausibility — and he stressed that the application of values may be considerably affected by individual factors such as personality and biography.

In my exposition of the conceptual process of science, the need for a disciplinary matrix — something which is common to a certain group of scientists — will become obvious. Whether in the sense of offering the theoretical background against which scientists gather data from systems in reality, or in the sense of forming research objectives and directing conceptualisations of real systems, or in guiding interpretations of theories in their application or implementation, a context somehow common to all scientists of a certain research group is indeed indicated.

Note though, that Kuhn (ibid., p.184) sees it as the function of models to provide the relevant community with preferred or permissible analogies and metaphors, and so to help to determine what will be accepted as an explanation and puzzle-solving solution, as well as to determine as yet unsolved puzzles and their relative importance to the group. In my terms, although models help to do this, all of these are functions of the matrix, rather than of individual models. Models are rather the result than the cause of these things and are thus not to be identified with the matrix itself. In my view, models have a specialised role in the sense that they are specific interpretations of given theories in which these theories are true. So, the formulation of models has to be preceded by factors such as determining the nature of these models, whether they will offer an explanation for whatever problem is being studied, and so on. The disciplinary matrix should in my terms be seen as offering the background against which actions at all three (model-theoretic) levels of the scientific process take place.

Traditionally it is thought that scientific knowledge is embedded in theory and rules. A student cannot learn to solve problems before she has learned the theory and the rules for applying it. And then, after having solved many problems, such a student will find it less and less difficult to solve more and more complicated problems. As we know, Kuhn (ibid., p.188) claims that

> ... at the start and for some time after, doing problems is learning *consequential things* about nature. In the absence of such exemplars, the laws and theories [the student] has previously learned would have little empirical content.

He further goes on to say (ibid., p.190) that scientists solve puzzles by modelling them on previous solutions. He wants to emphasise that what he refers to as "consequential knowledge of nature" is acquired while learning the similarity relationship and is afterwards embodied in a particular way of viewing nature — rather than by laws and rules.

He (ibid., p.191) claims the verbal statements of laws taken by themselves to be "virtually impotent"[4] and stressed that what actually results from the use of exemplars is something like Polanyi's "tacit knowledge" which is learned by doing science rather than by acquiring rules for doing it. Kuhn (ibid., p.194) argues that claiming that rules and the ability to apply them are acquired from exemplars implies that there are, already at that stage, alternatives in the following sense. We might, he (ibid.) continues, "... have disobeyed a rule or misapplied a criterion, or experimented with some other way of seeing. [And those] are just the sorts of thing we cannot do." I would qualify this remark by adding that those are things scientists cannot do until after they have started making observations and thus taking part in the scientific process. Then they start abstracting data from real systems, constructing conceptual models, and maybe in the end, formulating theories, always directed by factors like research and personal goals and working from within a specific disciplinary matrix.

The only way in which we can have scientific contact with the world (that is, with systems in reality) is through actions involving selection, abstraction, and generalisation, which are always executed within some theoretical framework or disciplinary matrix, and are always teleological in character in the sense that these abstractions are made in order to theorise eventually about a specific aspect of some real system relevant for certain context-specific reasons. So in this sense there are indeed no absolute rules and laws guiding us towards the expression of our knowledge, because of the way our encounters with reality are structured. And, in this sense, it is indeed only by constructing various models of reality that we learn "consequential things about nature".

Thus, the way in which models function in a model-theoretic interpretation of the process of science has some of the features that Kuhn ascribed to this second use of the notion of "paradigm". Models can be seen as the basis for solving "puzzles" in the sense that they are actually constructed in the first place, when they form part of the process towards theory-formulation, to do just that. Also, more importantly, when considering the implementation or application of theories, already established models of theories may lead scientists to new interpretations of these theories, or to amending the existing models in order to offer "better" explanations for the problem the theory is addressing. To claim however — as Kuhn does — that models replace explicit rules as basis for puzzle-solving is maybe taking matters a bit far. It is not possible to find a unique description dictating the content of constructed models (whether during the process towards theory-formulation, or afterwards, when the theory is applied), simply because these constructions are so context— and theory—specific. However, this is part of the changeable nature of scientific knowledge, and so, the fact that models operate as a link between theories and systems in reality, and that they have an essential interpreting role to fulfil in the process of science, is a significant feature of scientific practice.

As far as Kuhn's claim that models offer scientists a specific way of viewing nature is concerned, that is indeed the case in a model-theoretic interpretation of the scientific process too (although the broader motivation for a specific world view is taken to be offered by a specific disciplinary matrix). Both in the process of theory-formulation and during the application of theories, models are constructed and employed in accordance with a certain view of reality, and afterwards, these models exist in a sense as an affirmation of these views.

In the rest of this chapter, I shall concentrate on the nature of scientific theories, their relations to models and to (systems in) reality, and the implications thereof for the nature of scientific knowledge. I shall start off in Section 2.3 with a discussion of the formulation process of scientific theories, i.e. their "coming-into-being". Then, in Sections 2.4 and 2.5 I shall discuss the matter of applying and interpreting scientific theories. Finally, in Section 2.7, I shall conclude this chapter by looking at the problem of what I term "empirical proliferation", that is the over-determination of theories by their empirical models. The issues discussed in this chapter shall be picked up again and worked out finally in terms of their realist implications in Chapter 5.

2. TERMINOLOGICAL NOTE

In this chapter I shall take examples both from the natural sciences — mostly from physics and astronomy — and from philosophy of economics to support and illustrate my arguments. As mentioned in Chapter 1 an account of science such as the one that I shall set out in this chapter works very well for the natural sciences, because all three main aspects of such a model are simple and clearly portrayed in the natural sciences. As far as the aspects of reality studied go, an electron is a far simpler concept than a human being; the models employed at the "middle" level of my interpretation of science can often be mathematical in the natural sciences, which they cannot necessarily be in other sciences; and finally, the languages used in natural science can be formalised more easily, while other sciences often use full natural language which is tremendously more complex. I am convinced though that the model that I am proposing for the philosophical interpretation of science is also applicable to the social sciences, although in this case the stage of theory formulation and the interpretative stage of the scientific process may differ from that of the natural sciences as far as certain emphases on context dependency and other related issues are concerned. Note that Suppes has applied his set-theoretical account of science also for instance to psychology, while the structuralists have published extensively in philosophy of economics (although in an instrumentalist context).[5] Apart form referring to certain issues in philosophy of economics here and there, the scope of this text however does not allow me to go into these matters in any more detail.

Before proceeding, it is necessary to give at least an informal explanation of the notions of "theory" and "model" as I shall be using them. Let us choose a first-order predicate language, L, in which a deductive theory T is formulated. The only condition I set with regards to language L, is that it should be appropriate for formulating statements about mathematical structures. Let us say that theory T is the (deductively closed) set of all formulae which can be deduced from a consistent set (system) of axioms, Σ, in formal language L. Now, in this language, L, there will be — among other things[6] — an infinite, countable set of individual variables and a nonempty set of predicate letters. Then, a mathematician (or "scientist" for my purposes) may give meaning to these symbols used to formulate the sentences in theory T in language L, by constructing a certain mathematical structure, call it U, suitable to be described by the language L.[7]

An interpretation of language L will consist of a set over which we consider the individual variables to range, and (predicates or) relations defined on this set as interpretations of the predicate symbols in L. Thus as soon as every *n*-ary predicate symbol in language L is associated with an *n*-ary relation in structure U, we can say that this mathematical structure is an *interpretation* of the language L, and thus, by implication, of any sentence in L.[8] Note that of course, for *every other definition* of the domain of the mathematical structure and of the relations defined on it, one is confronted with *another interpretation* of the language. There are thus no "rigid designators" across interpretations.

A *model* of any formula such that every free occurrence of variables in it refers to an element in the domain of an interpretation of L by means of a specific valuation, will be an interpretation under which that formula is true by the specific valuation defined for its (the formula's) variables.[9] Now, a *sentence* is a formula with no free occurrence of variables. Thus the definition of the truth of a formula implies that a sentence will either be true under an interpretation by all possible valuations or false under all valuations. Hence for sentences we may speak of truth under an interpretation without mentioning valuations. And, a set of sentences is true under an interpretation if every sentence of that set is true under that interpretation. Thus a *model* of a theory (being a set of sentences in some formal language L) will be an interpretation under which that set of sentences (i.e. the theory) is true.[10]

Let us make these informal notions more precise by considering the following definitions:

Definition 2.2.1

Suppose we have a set Ind of constants and a set Pred consisting of predicate symbols and their arities.

The set **A** of atoms is the set of all strings of the form $P(t_1, t_2, ..., t_n)$

such that

- $(P,n) \in$ Pred and

- $t_1, t_2, ..., t_n \in$ Ind $\cup$ Var, where Var = $\{x_1, x_2, ...\}$ is the set of *variables*.

The *first-order language* L_A generated by **A** is the set of all *well-formed formulas* over **A**, where α is a well-formed formula over **A** if and only if

- α ∈ **A**, or

- α is of the form ¬β, where β is a well-formed formula over **A**, or

- α is of the form (β∗γ), where β and γ are well-formed formulas over **A** and ∗ ∈ {∧, ∨, →, ↔}, or
- α is of the form (∃x)β or of the form (∀x)β, where β is a well-formed formula over **A** and x ∈ Var.

Definition 2.2.2

Suppose we have a first-order language L_A, where **A** is induced by some sets Ind and Pred.

An *interpretation* of L_A is a pair (D,f) such that

- D ≠ ∅,
- f assigns to every member of Ind an element of D, and
- f assigns to every (P,n) ∈ Pred a subset of D^n.

If Pred contains the *equality* predicate constant (=,2), then we stipulate that f(=,2) must be the subset $\{(d,d) \mid d \in D\}$ of D^2.

A *term interpretation* is a pair (D,f) such that D = Ind and, for every c ∈ Ind, f(c) = c.

Definition 2.2.3

Suppose we have a first-order language L_A and an interpretation I = (D,f).

An *assignment of values to variables* (or just 'assignment' for short) is any function v from the set Var = $\{x_1, x_2, ... \}$ to the set D.

Definition 2.2.4

Suppose we have a first-order language L_A, an interpretation I = (D,f), and an assignment v from Var to D.

The *valuation* corresponding to I in the context of v is the function w from **A** to {T, F} determined as follows:

- for every $P(t_1, t_2, ... , t_n)$ in **A**, assign to $P(t_1, t_2, ... , t_n)$ the truth value T if and only if $(d_1, d_2, ... , d_n)$ ∈ f(P,n); where each d_i is either f(c), if t_i is the constant c, or is v(x) if t_i is the variable x.

Definition 2.2.5

Suppose I is an interpretation of a first-order language L_A and v is an assignment in I. Let w be the valuation corresponding to I and v.

The interpretation I *satisfies* the wf α in the *context* of v iff one of the following conditions holds:

- α is an atom and w(α) = T;
- α is ¬β and I does not satisfy β in the context of v;

- α is $\beta \wedge \gamma$ and I satisfies both β and γ in the context of v;
- α is $\beta \vee \gamma$ and I satisfies at least one of β and γ in the context of v;
- α is $\beta \rightarrow \gamma$ and I satisfies γ or I does not satisfy β or both in the context of v;
- α is $\beta \leftrightarrow \gamma$ and either I satisfies both β and γ in the context of v or else I satisfies neither β nor γ in the context of v;
- α is $(\exists x_i)\beta$ and I satisfies β in the context of at least one assignment v' that differs from v at most on variable x_i;
- α is $(\forall x_i)\beta$ and I satisfies β in the context of every assignment v' that differs from v at most on variable x_i.

<u>Definition 2.2.6</u>

Suppose we have a first-order language L_A.

By W_A we understand the set of all interpretations of L_A.

An interpretation I that satisfies a wf α in all contexts is a *model* of α, and the set of all models of α is indicated by Mod(α).

More generally, if Σ is a set of wfs and I is an interpretation that satisfies each wf α belonging to Σ in all contexts, then I is a model of the set Σ; the set of all models of Σ is indicated by Mod(Σ).

A subset F of W_A is a called a *frame* and we say that F is *axiomatised* by a set Σ of wfs iff F = Mod(Σ).

A wf α *entails* a wf β iff Mod(α) $\subseteq$ Mod(β), and wfs α and β are *semantically equivalent* iff Mod(α) = Mod(β). (As before, we may write $\alpha \vDash \beta$ as abbreviation for 'α entails β' and $\alpha \equiv \beta$ to abbreviate 'α is semantically equivalent to β'.)

<u>Definition 2.2.7</u>

The variable x *occurs free* in the wf α iff one of the following is the case:

- α is an atom and x occurs in the string α (i.e. α has the form $P(t_1, \ldots, t_n)$ and x is one of the t_i);
- α is $\neg\beta$ and x occurs free in β;
- α is $\beta * \gamma$ and x occurs free in β or γ or both, where $* \in \{\wedge, \vee, \rightarrow, \leftrightarrow\}$;
- α is $(\exists x_i)\beta$ and x occurs free in β and $x \neq x_i$;
- α is $(\forall x_i)\beta$ and x occurs free in β and $x \neq x_i$.

The wf α is a *sentence* iff no variable occurs free in α.

In Section 2.7 of this chapter I shall discuss an application of non-monotonic logic to my model-theoretic account of science. Here follows a few of the main formal definitions that I shall refer to.

Definition 2.2.8

Let G be any finite set of possible worlds. A relation R ⊆ G×G is a *total preorder* on G iff

- R is *reflexive* on G (i.e. for every x∈G, (x,x) ∈ R), and
- R is *transitive* (i.e. if (x,y) ∈ R and (y,z) ∈ R, then (x,z) ∈ R), and
- R is *total* on G (i.e. for every x∈G and y ∈ G, either (x,y) ∈ R or else (y,x) ∈ R.

Definition 2.2.9

Let L be a propositional language over some finite set **A** of atoms. Let **W** be the set of all local valuations of L (i.e. functions from **A** to {T, F}). A *ranked finite model* of L is a triple M = (G, R, V) such that

- G is a finite set of possible worlds,
- R is a total preorder on G, and
- V is a labelling function from G to **W**.

By a *default* model of L we understand a ranked finite model (G, R, V) in which G = **W**, R is a total preorder on **W**, and V is the identity function (i.e. V(w) = w for all w∈**W**).

Definition 2.2.10

Suppose that L is a propositional language over a finite set **A** of atoms, and that M = (G, R, V) is a ranked finite model of L. Given a sentence α of L and a possible world x ∈ G, the following rules determine whether M satisfies α at x:

- if α is an atom in **A**, then M satisfies α at x iff the valuation V(x) assigns to α the truth value T;
- if α is ¬β then M satisfies α at x iff M does not satisfy β at x;
- if α is β∧γ then M satisfies α at x iff M satisfies both β and γ at x;
- if α is β∨γ then M satisfies α at x iff M satisfies β at x or γ at x;
- if α is β→γ then M satisfies α at x iff M satisfies ¬β at x or satisfies γ at x;
- if α is β↔γ then M satisfies α at x iff M satisfies both β and γ at x or satisfies neither at x.

Definition 2.2.11

Suppose L is a propositional language over a finite set **A** of atoms, and that M = (G, R, V) is a ranked finite model of L. Let α and β be any sentences of L. The sentence α *defeasibly entails* β iff M satisfies β at every possible world x such that

- M satisfies α at x, and

- x is minimal amongst the worlds satisfying α, i.e. there is no possible world y of M such that α is satisfied at y and (y,x) ∈ R and (x,y) ∉ R.

3. THE FORMULATION OF SCIENTIFIC THEORIES

Now, turning to the epistemic process leading to the formulation of a scientific theory, the following. No-one — not even scientists — ever studies reality in all its complexity. The way in which we come to (scientific) knowledge is determined by acts of abstraction and simplification. Thus, rather than focussing on the colourful richness of reality, scientists typically will decide to focus on a particular aspect or system of reality. Moreover, intensifying their initial selective actions, scientists will also decide to concentrate only on particular features of the real system they have picked out.

At the start of a particular line of research, the first encounters between scientists and the relevant system in reality have an interesting feature. Although traditionally viewed as happening at the lowest level of scientific activity — "lowest" in the sense of least abstract and not least dependent on historical, social, and cultural factors — these encounters are already not "objective" in the sense of being neutral to any kind of external influence. This is because of the influence various contingent factors have on the actions of scientists and their arguments. These factors range from extremely specific to broad combinations of general factors influencing scientists at a given time. They include personal factors such as the personal interests of scientists involved, their particular research goals, and the social context in which their research is done.

Then there are also factors that pertain more to the "theory-ladenness" of the choice of experiments, the interpretations of data, and so on. These factors include the paradigm or research tradition in which the scientists are working, the state of their discipline at the time, the level of technology and experimental apparatus available at the time to those particular scientists, and also the body of "already established" theories (as the background) against which these particular scientists will work. Also included here are factors to which Gerald Holton (1995) refers as "themata". These factors are a cross between, on the one hand, the specific motivation behind the choice of addressing results and problems within one particular scientific framework rather than another, and, on the other hand, the scientists' world view at the time.[11] In this sense, I agree with the constructivists: no scientific activity takes place in some kind of objective vacuum. All the above factors form part of at least the relevant disciplinary matrices, and usually also influence choices of models of theories, or even model construction.

The inherent conditioning, or refining, abstractive goal (and potentially idealising power) of activities carried out at this level leads ensuing activities to a level more general in scope than that of the original encounters with specific aspects of reality. At this level scientists create conceptual "models" — which I call "intended models" for obvious reasons — of the real system in question. These models are obviously not (yet) formally identifiable as interpretations of any sentences of the language in which the final theory will be formulated. However, after theory-formulation, at the stage where possible interpretations and (empirical) applications of the theory in question are considered, it will become clear that the intended "model" of the theory in question is also one of the (possible) mathematical models (i.e. interpretations under which the theory in question is true) of the relevant theory.

Let us consider briefly the formulation of Newton's laws of motion and his law of gravitation. Newton wanted — inter alia — to continue Kepler's research about the movement — and positioning — of the planets in *our* solar system. Kepler's laws originated, it seems, largely because of his own interest in specifically the movement and positions of the planets in our solar system. His intended model thus may be said to have been supported by data concerning only these (and related) planetary features. His research was based on observational data regarding the positioning of the planets at different times, much of which was the original work of Tycho Brahe. His interpretation of his observational data would have been, for instance — and probably among other things — influenced or "laden" by his mathematical idealism — e.g. his claim that the planetary orbits should fit exactly into nested Platonic solids. Finally, Kepler's research culminated in his formulation of his three laws — the first two in 1609 and the third in 1618.

Now Newton could not study all the complexities of our solar system as it manifests itself in the manifold of reality. He was also, as mentioned above, interested in examining planetary motion in our solar system. Thus he identified the details necessary for his research goal by abstracting from this system in reality only those specific features in which he was interested. But, in this way, because these abstractions were so closely guided by his intentions — and certainly influenced by both Brahe and Kepler, and also Galileo's findings — he would never really have been dealing with the "bare" data that he had extracted from reality. He would, rather, in fact, have been dealing with a conceptualised model of our solar system that would in the end lead him to the formalisation of the theory of solar systems itself.

In order to study the dependence of the force of gravity on the distance from the centre of the earth, Newton compared the fall of a stone on the surface of the earth with the motion of the moon. Newton discovered that the "... forces of terrestrial gravity decrease as the inverse square of the distance from the centre of the earth" (Gamov, 1962, p.62). He consequently generalised this result to "all material bodies in the universe" (ibid.) and so formulated his universal law of gravitation.

Scientists thus conceptualise their objectives in the light of the data they gather — and may still be gathering — with an eye on their research goals and guided by the specific scientific tradition, community, theoretical network or paradigm they are working from. The creative context of this stage of the scientific process offers scientists the chance to test and constantly reformulate their conceptual structures and to receive results under the conditions set by their goals and the context within which they work. This implies that these (intended) models have an idealised nature in the sense that they are the results of extremely focussed actions which typically disregard factors in the empirical system in question that could muddy the waters of their research.[12]

The progression of generalisation common to this stage of science may perhaps roughly be logically reconstructed to range[13] from

- scientists' initial sensations (possibly mediated by some apparatus) of "real" objects and their behaviour in some real system (broadly, of aspects of reality), to
- the construction of percepts of these sensations, to
- the construction of concepts of these percepts, to
- the construction of conceptual models which are structured sets of these concepts, and which
- may then — in certain sciences — culminate in the formulation of mathematical models.

Finally, these (abstracting) actions culminate[14] in the formulation of a general (abstract) theory — expressed (or expressible, at least) in some suitable language — in the field of research in which the relevant scientists have been working. The nature of this level at which theories are formalised is abstract, general, and simple in the sense that the values (meanings) of the parameters in the general theories are essentially *unconditioned* and the meaning of theoretical terms (such as "electron" or "mass") is in principle *open* to valuations or interpretations made by scientists interested in applying or implementing the theory. This implies naturally that a potentially infinite number of conceptual interpretative (or mathematical) models can be constructed of one and the same theory.

The aim of Kepler's research surely was to formulate some kind of law (or laws) concerning planetary motion. It is sometimes claimed that Kepler's laws do not constitute a scientific theory, however. For instance Dilworth (1994, p.135) claims:

> The main reason usually given for Kepler's laws, taken together, not ranking as a theory is their unequivocally empirical character — i.e. the fact that they are 'instantiated' in the sense that they refer to the individual planets in the solar system and, unlike a theory, are capable of being tested more or less directly. The present view [Dilworth's] supports a distinction along these lines, and in fact provides an explanation of it, viz., that, unlike Newton's theory, Kepler's laws are not integrally related to a model ... .

Well, it is indeed the case that Kepler's laws are very "empirical" and less general in scope than Newton's laws of motion and his law of gravity, since the latter do not give any particular value or parameter to their theoretical terms.[15]

However, Kepler does generalise from his intended model in the sense that his laws are about the motion and position of all the planets — or any planet — (known to him) in our solar system, and not particularly about any one specific planet. Moreover, the application of his laws to the motion and position of any one planet may be seen to imply the construction of a specific conceptual model concentrating only on the particulars concerning that specific planet. Although his laws are indeed far less general than Newton's, and also far more like the kind of empirical "theory" philosophers in the Popperian tradition advocate, his laws can be viewed as part of an "intellectual system" of the kind Torretti (1990, p.24) supports. Torretti's argument comes down to the following: Should it be found (as perhaps from a certain perspective it was discovered much later) that a particular planet does not obey Kepler's laws it would imply a revision of our scientific thoughts concerning planetary motion. I also take this as sufficient motivation to view Kepler's laws as comprising a theory (albeit perhaps a "low-level" one).

It is the case that Newton's laws of motion and his law of gravity could in fact finally *explain* Kepler's laws, and thus that Kepler's laws may perhaps not be said to *explain* the positioning of the planets but rather merely to *describe* their motion. I think that we have at least to accept the legacy of the advocates of the deductive-nomological model of explanation in so far as we accept that scientific explanation is really some kind of inference, the conclusion of which *describes* the facts to be explained. Generally, Newton's laws *explain* Kepler's in the *model-theoretic* sense that an interpretative model of Newton's laws may be given by our solar system, and the data giving positions and motions of a particular planet in that solar system may be viewed to constitute an empirical model that is isomorphically embedded into the above interpretative model.

More formally, the application of Newton's universal law of gravity to the motion of our solar system's planets around the sun (interpretative model) enables one to mathematically derive Kepler's three laws. More specifically for instance, Newton's law of gravitation explains Kepler's third law in so far as it shows that Kepler's third law is based on a force that is exerted towards the centre of the sun and that is inversely proportional to the square of the distance between the sun and the planet in question. However, it has to be noted that although Newton's laws "explain" Kepler's (in the above sense) they do *not*, after all, explain e.g. gravity — they merely describe it. (I shall briefly discuss the explanation-description debate again in Cartwright's terms in Chapter 4. This, though, is a very complex debate that cannot be fully analysed within the scope of this book.)

Do Newton's laws constitute a theory? Should his laws be viewed as the axioms of this theory or as empirical laws in the Popperian way? In the model-theoretic approach to science that I advocate, the axioms of a theory describe the interpretative model(s) in which they (or the laws they represent) are true. Such a model then sets out the calculation of values of certain functions and the interpretation of certain theoretical terms in the context of the model. In this case, for instance, it could mean that bodies are conceived of as "mass-points" without extension. The theory in general sets down the nature of the relations between the terms in its interpretative models. Thus in a model-theoretic realist context, Newton's laws of motion and his law of gravity do constitute a theory since they are general enough, or broad enough in scope to offer different interpretative models or interpretations of its terms in different contexts of application.[16] They can be viewed as "empirical" laws only via the interpreting mediation of interpretative models and some empirical models representing observational data and other empirical calculations, which brings us to the next section on the interpretation of scientific theories.

4. THE PRIMARY SEMANTIC INTERPRETATION OF SCIENTIFIC THEORIES

Now, in order for scientists to give "reference" to the multi-interpretable theoretical terms and parameters[17] of scientific theories, more models — other than the original one leading to the formulation of the theory — may be constructed. In other words, because of the possibility of all these different models of theory T (in language L), the theory — say in our example, Newton's laws — may be (in principle at least) related to any (mathematical and thus conceptual and interpretative in my terms) model in which it is true and not only to the intended (conceptual) one. If we take the set of axioms, Σ, to be Newton's three laws of motion and his law of gravitation, expressed formally, then they will hold in any planetary system (because of the general nature in which they were formulated) and then we can pick any "planet" in such a system and be sure that its orbit will be an ellipse and the system's "sun" will be in one focus. E.g. a model in which, say, Jupiter is the "sun" and its satellites are the "planets" may now be constructed.

These models are thus interpretations of the theory[18], each of which is also in its turn determined by — among other factors — the research intentions and thematic preferences of the scientists wishing to apply or study the theory. Note here that the first most obvious model of the theory is its original "intended" one. However, since different groups of scientists will be applying the theory in question — perhaps for different reasons — at various times, this is not necessarily the model that will be chosen as the one via which the theory is to be applied or interpreted. As we shall see in Chapters 3 and 4, the problem of defining an "intended" model, or a class of "intended models" cause both statement and non-statement supporters a lot of problems. In what follows I shall claim that the fact that these "intended" models cannot be uniquely "bounded off", as it were, does not necessarily have anti-realist implications though,especially given the articulative possibilities of reference relations that model-theoretic realism offers. Other problems with defining these kinds of model, such as Putnam's model-theoretic paradox, shall be examined in Chapter 3.

The intended model is thus in nature no different from the mathematical structures that will be constructed to interpret the theory in such a way that the theory will be true in them. These models simply differ as far as the nature of their origins is concerned, and features common to both — such as the role of thematic preferences — are simply emphasised differently in each case. In this sense for example, the intended models have more of an organising and guiding role in the sense of being the first conceptual means via which scientists are able to make the first abstractions from reality. The interpretative models, in their turn, will give reference and meaning to or "fill in" the content of (some of) the general terms (e.g. electron, mass, velocity, temperature) used in the theory[19] and specify values for the parameters of formulae in the theory in such a way that the theory turns out to be true in these models. The models thus constructed are then obviously mathematical models in the Tarskian sense.[20]

In the following I shall limit the discussion to model theory of one type of language, namely first-order predicate languages (see eg. Heidema, 1972). The set-theoretic modelling of syntax and semantics comes a long way, at least since the early

Tarski in the early 1930's. Model theory is the mathematical, or set-theoretical, study of the interplay between syntax and semantics. "Historically it has its roots in the various attempts of reducing ... mathematics to logic (Frege, Hilbert), ... logic to number theory (Skolem, Gödel), and finally, of modelling logic within set theory (Tarski, Vaught)." (Makowsky, 1994, p.241).

In order to introduce the difference between non-statement, and statement accounts of science, and to show where a model-theoretic account fits in, I shall use the distinction between interpreted and uninterpreted languages that Marion Przelecki (1991) offers in an article entitled "Is the notion of truth applicable to scientific theories?". According to this distinction we have a (suitable first-order) language L characterised only syntactically ("by rules of formation" (ibid., p.285)), and an interpreted language L' characterised also semantically (according to "rules of interpretation" (ibid.)) and identified with a pair <L, I> with I an interpretation of L.

In the conceptual framework of standard model-theoretic semantics this translates into the following: The interpretation I of language L is identified with a "suitable set-theoretic structure", called a model M of theory T in L. The relative notion of truth for L is in this context represented by the claim that sentence α in L is true in model M. In a specific context, concentrating on a specific real system, it is claimed then, that there is, among the possible models of T, at least one __M*__, corresponding to the actual or intended application of T, which is an actual or intended model of T in L. This implies that the notion of truth is directly applicable to scientific theories only if they consist of statements, as identified with interpreted sentences. (ibid.).

This distinction seems to correspond to the distinction between statement and non-statement depictions of science. The term "statement" however in the non-statement account, is not distinguished clearly from the term "sentence", since the defenders of the non-statement account maintain that scientific theories are not composed of sentences, actually they are said not to be linguistic entities in any way. Accordingly, as Przelecki (ibid.) remarks, the statement view is "meant to involve all conceptions which treat scientific theories as some sets of sentences ...". In this sense, my model-theoretic proposal is then a "statement"-oriented proposal, although the current issue under consideration, i.e. the interpretative non-linguistic models of theories, is central to my view, which would then make it a "non-statement"-oriented proposal. This shows, as mentioned before, that the difference between these two approaches is far more one of emphasis than an actual one.

Also, in a model-theoretic account of science you will see that reference to systems in reality are claimed to be traceable if there is at least one so-called "empirical model" that stands in a certain kind of relation to a special subset (called an "empirical reduct") of an interpretative model of the relevant theory. However nowhere is it implied that there is either only one interpretative model of a certain theory, or only one empirical model with these special features. Indeed, there are many of both. This issue will be discussed in Section 2.7.

A last point in this regard: finding certain models standing in these relations to each other does not depend on them being "intended" models. As mentioned often now, the multi-interpretability of formal languages by their non-linguistic models implies that theories are applicable to many more models other than just one specific "intended" one — without any negative implications for tracing realist references I might add. A scientist has, in effect, very little control over the possible models of her theory, in the sense that she cannot be expected to know about, or take into account, before formulating her theory, all the different scientists who will, in time, become interested enough in her theory to construct their own models thereof. We all know that Einstein, for instance, had a static universe as his intended model for his general theory of relativity. It so happened, however, that other interpreter-scientists of Einstein's theory, constructed models in which the universe was anything but static. All that was left for Einstein to prevent such models being constructed, was to change his original set of assumptions. In other words, the only sense in which the scientist has control over the construction of a multiplicity of models of her theory, is in a reactive sense, namely that she can go back to her original set of assumptions and change them in such a way that the unwanted model is not a model of this set any more.[21] Again, more on this in Section 2.7.

Returning to our discussion of interpretative models of theories, although more specific than the theory, these models are still general in nature in so far as they are idealisations in the same way as the intended models are. Nancy Cartwright (for instance Cartwright, (1983), (1986), (1989)) sees this as problematic as far as the possibility of theories offering descriptions of reality is concerned. I disagree, as will be discussed in Chapter 4, and as will become more apparent in what follows. Although we may model-theoretically define the *form* of links of empirical adequacy, it is impossible to give clear-cut rules or conditions for the *content* of the adequacy of our conceptions to the "real" existing objects in systems in reality, because of the *open-ended* generality of theories in the sense of the different models in which they may be true and which is a consequence of the "abstract" character of theories and also because of the open-endedness of the interpretative models in the sense of their "ideal" nature. Establishing the last referential link between some system in reality and some model of the theory forces a suspension at the conceptual level of the *ceteris paribus* clauses at play at the linguistic level of science in order to fill in the details that, at this level, have been "idealised" and "abstracted" away. More about this in what follows, and also in Chapters 4 and 5.

Here is a simple and familiar example to illustrate the construction of different models interpreting the same theory. Newton's laws made it possible to calculate very precisely the motion of the planets in our solar system (any solar system for that matter) under the influence of mutual gravitational attraction. Up to 1820, scientists interpreting (or applying) Newton's laws of motion and his law of gravitation to our solar system had worked in a model of these laws (comprising Newton's "theory") which consisted of only seven planets. Then, in 1820, calculations carried out within this model started to give "wrong" predictions, and it became apparent that the motion of Uranus "did not conform to Newton's grand scheme" (Schwinger, 1986, p.195). The possibility that the motion of Uranus could be affected by the gravitational attraction of another planet seemed a good solution to the problem though. So, scientists thought of postulating the existence of an eighth planet, and consequently constructed a different model of Newton's theory, now with eight planets. In 1845 John Adams calculated the position of this "new" planet — Neptune — in our solar system, and shortly afterwards Urbain Leverrier's calculations confirmed Adams's findings.

Applications of the same Newtonian "theory" (his three laws of motion plus his law of gravity) includes the "discovery" of Pluto in 1930 as the result of theoretical calculations based on the universal law of gravity. Also Newton gave the first explanation of the precession of the equinoxes since the time of the Greeks by applying his law of gravity to the motion of the earth. And, a last example, aspects of the motion of the tides of the sea could be explained by applying the universal law of gravity to the earth's perihelion and aphelion motions (i.e. the movement of the earth far from and close to the sun). Thus new information results in different (new) models still constructed to attain the same (previous) goal, but also different aims result in different models.

I take the relations that exist between some theory and the mathematical (interpretative, conceptual, semantic) models that are interpretations of the theory's language and in which the sentences of the theory are true, as the first set of relations that determine the possibility of reference to some real system. We need, however, a second set of (much more complex as it turn out) relations to refer us to (aspects of) reality. The next section focusses on this empirical interpretation of theories.

5. THE EMPIRICAL INTERPRETATION OF SCIENTIFIC THEORIES

The goal of a formal logician will merely be to prove that her deduction (theory) is valid, i.e. true in all possible worlds allowed by the axioms, i.e. true in all possible models (in the conceptual system), one of which may or may not be in its turn "about" some system in reality. Thus, for the formal logician, the question of whether it is possible to construct a "second set" of interpretations or models (to retrace the steps of the original scientist — representing the group of scientists that "formulated" the theory in question — even further back to reality), is rather irrelevant.

Scientists, however, surely will be interested to know whether one of the interpretative models of their theory can have a system in reality as some further interpretation or model, because they formulated their theory precisely to enable them to make some sort of claim about a certain aspect of some real system. The method of verification of each of these (interpretative, mathematical) models (i.e. how well do each of them reflect the system in the real world?), will be decided by the specific nature of the specific (interpretative, mathematical) model in question, *as well as* by the nature of the specific real system in question. It could be that an observation through a telescope is needed, or an observation through a microscope, or some sort of calculation, which has less to do with observation, and so on and so on. In other words, neither Tarski, nor anyone else, could or can really specify the content of a general criterion for the truth of the sentences in this last set of interpretations.

Please see *Figure 1* below for what follows.

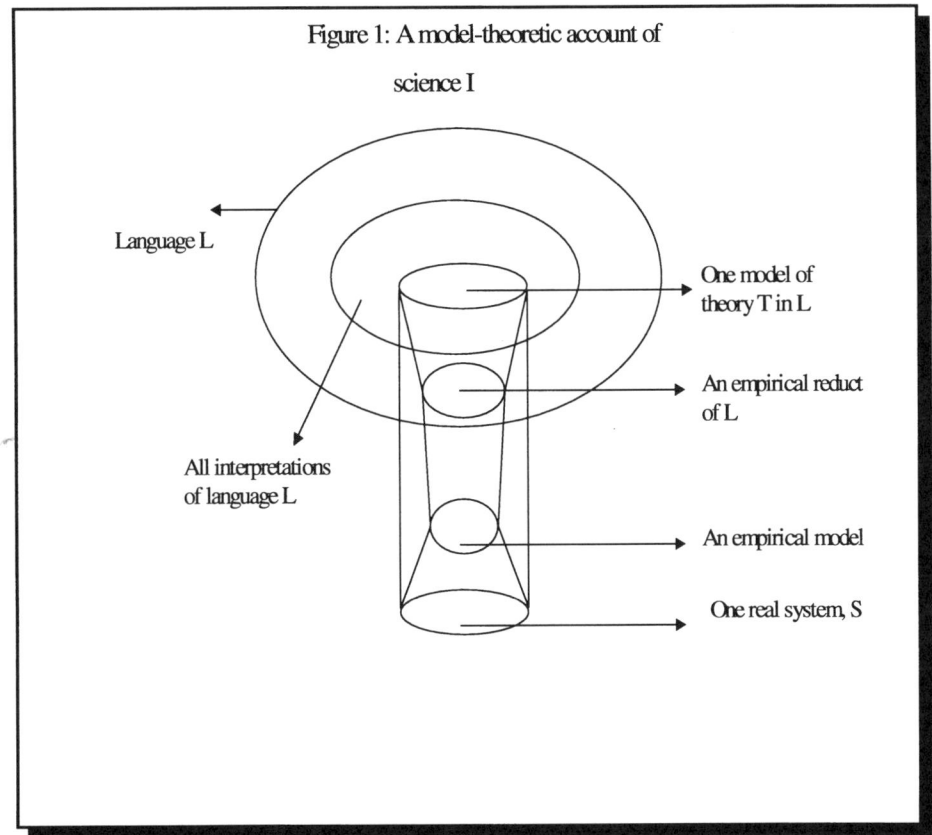

Figure 1: A model-theoretic account of science I

Hence I claim that if the phenomena in some real system and the experimental data concerned with those phenomena are logically reconstructed in terms of a mathematical structure — call it an "empirical" model — the relation of empirical adequacy then becomes — close to Van Fraassen's depiction — a relation which is an isomorphism from the empirical model into some empirical reduct of the relevant model of the theory in question.

Consider what it really means to formulate a model of a particular theory. A model of a theory sees to it that every predicate of the language of the theory has a definitive extension in the underlying domain of the model. Now, focussing on a particular real system at issue in the context of applying a theory, which in its turn implies a specific empirical set-up in terms of the measurable quantities of that particular real system, it makes sense to concentrate only on the predicates in the mathematical model of the theory under consideration that may be termed "empirical" predicates (in the particular context of application). This is how an empirical reduct is formulated. Recall that a "reduct" in model-theoretic terms is created by leaving out in the language and its interpretations some of the relations and functions originally contained in these entities. This kind of structure thus has the same domain as the model in question but contains only the extensions of the empirical predicates of the model. Notice that these extensions may be infinite since they still are the full extensions of the predicates in question. An empirical model embedded into an empirical reduct (of a model of a theory) usually involves only finitely many elements from the domain of the model (or reduct) and from the extensions of the predicates.

Now, from the experimental activities carried out in relation to the real system we are focussing on, a conceptualisation of the results of these activities, i.e. of the data resulting from certain interactions with this system, may be formulated. This (mathematical) conceptualisation of data is referred to as an empirical model. Then, if it is the case that there exists some relation of reference between our original theory and the real system we are considering, we may then find that there is a one-to-one embedding function from the empirical model into the empirical reduct in question.[22] Why? The empirical model contains finite extensions of the empirical predicates at issue in the empirical reduct, since only a finite number of observations can be made at a certain time.

To summarise: The interpretative model interprets all terms in the appropriate relevant language and satisfies the theory at issue. In the empirical reduct are interpreted only the terms called "empirical" in the particular relevant context of application or empirical situation. Think of this substructure of the interpretative model as representing the set of all atomic sentences expressible in the particular empirical terminology true in the model. An empirical model — still a mathematical structure — can be represented as a finite subset of these sentences, and contains empirical data formulated in the relevant language of the theory. See *Figure 2* for the example following below.

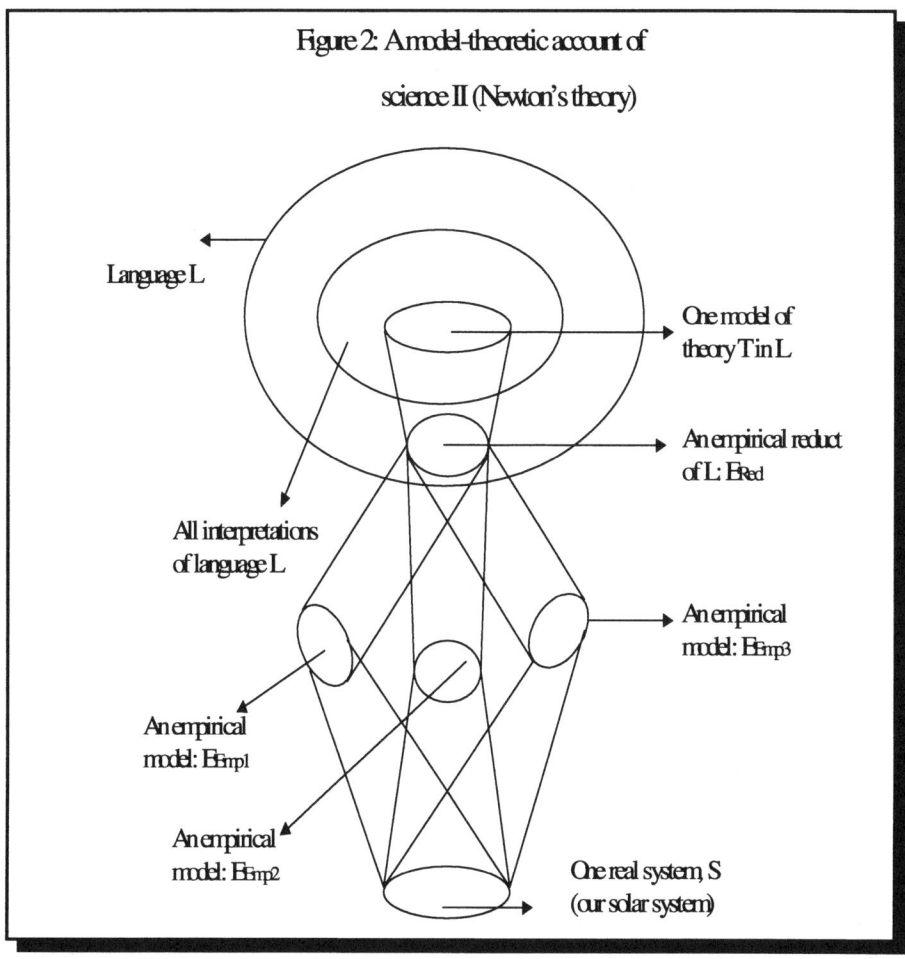

Say we take Newtonian mechanics as our theory. Take our solar system as a model, M, of the theory. Take one empirical reduct of this model, call it E_{Red}, a subset of M, containing as elements only events, that is, four-tuples (x,y,z,t) pinpointing the position(s) of Mars on its elliptical orbit. Notice that we acknowledge that the elliptical form of the orbit is an approximation, since we assume for now that the sun is heavier than any of the other planets and that we exclude predicates concerning forces, accelerations, and other so-called theoretical predicates — such as Mass — which are not the "direct" result of observations in this case[23]. This subset E_{Red} then is the set of all points (x,y,z,t) lying mathematically on the elliptical orbit of Mars. Should we now consider the empirical models that resulted from observations of countless astronomers through the ages, we would find empirical models E_{emp}^i, $\forall i \in \mathbb{N}$ all isomorpically embedded into our empirical reduct E_{Red} (assuming for our purposes here that Mars's orbit has not shifted for any reason). Thus we find that the conceptual four-tuples we get from observing the positions of Mars in space and time, that is, the elements of some empirical model E_{emp}, are amongst the elements of E_{Red}, that is, the four-tuples (x,y,z,t) showing us the position of Mars at various time instances.

Note thus that the referential relations between the theory and its models are much more simple — although also neither rigid nor absolutely fixable — than the relations between interpretative models and empirical models of systems in reality. The latter are extremely complex and never passive (or absolute), because in this case there are so many more variable factors to take into account when considering this more informal and supple relation. I quote Sir Allan Cook's (1994, p.141) explanation of the link between observation, models and theory in physics to show this more clearly:

> Observation is never an isolated activity. The way that we observe depends on human capabilities and properties of nature. Observation may affect the objects observed and our observational procedures depend upon the state of technology and are guided by theory. The results of observation [represented by my empirical models] have to be derived by procedures that depend upon some theoretical model [an interpretative model] as well as upon experimental techniques, ... The harder we question nature, [and] the more fundamental the observations we make, the more dependent are the results on technique and theory.

Moreover we as philosophers cannot tell — especially not "before the fact" — which specific conceptual construct (which interpretation of some theory) provides the most adequate description of some relevant system in reality. Only science itself can offer us — at some more mature stage of scientific development — an ontology (or ontologies) which can specify the *contents* or *detail* of the structures reality contains and the particular ways in which they behave.[24] Thus, neither the adequacy ("truth") of our conceptions nor the "reality" of the system as described by some theory, is absolute, because both are products of epistemically relative interpretations and subject to change. "Adequate" scientific statements may, indeed, say of reality that it is the way it is. It is, however, only through the mediation of (interpretative and empirical) models that this can be established, and never directly by somehow comparing *theories* to reality. More about this in Chapter 5.

A last comment on the relation between real systems and empirical models. It might seem as if real systems are in a sense in a model-theoretic context identifiable with sets of empirical data. This might be the case because it seems the closest we can "get to" reality in model-theoretic realist terms, is to find the isomorphic embedding function from empirical models into reducts. Taking also into account that an empirical model might conceptualise a set of data that is only concerned with an aspect of some real system, perhaps we should be satisfied with saying that different aspects of real systems may be reflected by different sets of data which may be conceptualised by diffferent empirical models.

But do unobservables "exist" in real systems? Well, model-theoretic realism *does* imply that the terms in theories refer to objects or relations in systems in reality. Recall that the "empirical" nature of reducts is contingent on a certain interpretation and empirical situation. Thus saying that "electrons" exist without any reference to models or interpretations or reducts is simply not really sensible. Recall also the emphasis on the re-interpretability of the language of science, or of theories in particular, and then it will be clear that claiming *model-theoretic reference* is sufficient to establish a form of realism, since in this *referential semantic* sense it can be shown that unobervables "exist" in real systems (i.e. terms in theories might after all be shown to refer to them). The contextually empirical terms refer directly, and the contextually theoretical terms indirectly, "by implication", via their conceptual and logical links to the empirical terms established by the theory. Some philosophers might be scornful about this kind of "weak" realism, while actually this realism is "weak" only because "strong" means traditional metaphysical realism. "Weak" means non-absolutist, and in that sense model-theoretic realism is much stronger and more flexible than typical metaphysical scientific realism.

6. THEORIES AND MODELS IN METHODOLOGICAL STUDIES OF ECONOMICS

Let us now examine the different components of the economic process a little more closely in model-theoretic terms. Let's agree for the sake of argument that economic theories may be reconstructed as generalisations (or sets of generalisations) concerning the (causal in transcendental realist terms) mechanisms underlying economic phenomena. Just as in the natural sciences, a "theory" may then be logically reconstructed as some linguistic expression containing some generalised description of a certain section of the system in reality being studied, the aim for the formulation of which is to enable one to answer questions regarding predictions, explanations, and descriptions of the phenomena involved.

Let's take the assumptions concerning a perfect market, i.e. complete or full information, the presence of many competitors, and the homogeneity of goods, as the assumptions applicable for a supply and demand theory (partial equilibrium). Let us add the following three axioms (in the structuralist empirical law sense): When, at the price ruling, demand exceeds supply, the price tends to rise. Conversely when supply exceeds demand the price tends to fall. A rise in price tends, sooner or later, to decrease quantity demanded and to increase quantity supplied. Conversely a fall in price tends, sooner or later, to increase quantity demanded and to decrease quantity supplied. And finally, price tends to the level at which demand is equal to supply. Let's take the definition of a demand function as the definite relationship that exists at any time between the market price of a good and the quantity demanded of that good. Let's take, in the same way, the definition for a supply function as the definite relationship that exists at any one time between market prices and the amounts of the good that producers are willing to supply.[25]

So-called "theoretical"[26] models — which are at the level of model-theoretic interpretative (conceptual, mathematical) models — in economics are "specific examples of theories" (Backhouse, 1997, p.159). In model-theoretic terms, of course, a model is a mathematical structure making the statements of the theory in question (being applied at a given time) true. Thus, we can take the simplest model of price theory, namely a linear model[27] and in these terms we can then draw the demand and supply curves as linear graphs and determine the equilibrium price as the point of intersection of the two graphs.

Shifts along the curves (i.e. changes in the quantity demanded or supplied in response to a change in price), or shifts of the curves (the results of changes in exogenous variables, i.e. changes in any of the factors held constant in the *ceteris paribus* clauses), can then be interpreted graphically via this linear model. The assumptions (auxiliary assumptions in economic language) needed to decide on the form of the function in question (i.e. linear or non-linear in this case), as well as on which variables to focus on and which to hold constant, are also part of the construction of the model, although at this stage these assumptions are still very general in nature.[28]

Empirical models are more specific. Remember that in model-theoretic terms, these models are isomorphically related to empirical reducts of the mathematical model in question, and may be viewed as representing subsets of the (empirical) sentences represented by the relevant reducts. So, via some further auxiliary assumptions, the forms of functions may be specified in detail, variables may be defined in terms of measurable observed parameters, and coefficients may be quantified using statistical (or other) methods. In our S&D example, we can now study a linear model of a specific market for instance, and as soon as data concerning exogenous variables are given to us, we can use our empirical model to make quantitative predictions, which opens the possibility of examining the consistency of our interpretative model with empirical data via the relevant empirical model(s).

Given the applied nature of economics, empirical models as conceptualisations of empirical data are important to realist economic methodology. Gibbard and Varian (1978) talk of "applied" models that result from theoretical modelling by interpreting the variables and general terms in the theoretical model in terms of some particular real world situation. Hausman (1992, p.76) claims that

> If the predicates models define have an extension — if actual systems satisfy these definitions — then with the proper interpretation of some 'theory' in the logical positivist sense — that is, of some set of sentences — the actual systems will be 'models' in the positivist's sense of those sentences.

This is in line, in the model-theoretic sense, with empirical models and the relations between empirical reducts of interpretative models and empirical models, rather than merely with interpretative models as interpretations of theoretical generalisations.

Connecting the different levels of scientific conceptualisation remains the challenge. The conceptual positioning of models, i.e. between origin and destination, between idealised (abstracted) theory and "raw" data, renders them invaluable critical mediators between science and reality and so helps to address this challenge. This is what Suppes (1962) referred to when he claimed that models are not of theories or of data or of phenomena, but rather somewhere in the middle, incorporating different degrees of both theories and data (see Chapter 4).

Mary Morgan (in De Marchi, 1988, p.199) points out that econometricians have always been occupied with finding "applied counterparts to theory that 'worked' with reference to observed data" (ibid.). She (and econometricians in general) might however have a slightly different interpretation of the notion of "empirical model". For instance she (ibid., p.200) remarks that the original goals of econometricians were to make economic theories more "concrete" and to measure the constant parameters of the laws of these theories, and since the "theoretical models" (which can either refer to the theory or an interpretative model of the theory in my terms, since economists seem to use these notions sometimes interchangeably) were not measurable, finding empirical models that could be subjected to measurement became necessary. In my terms empirical models are the means via which the mathematical or statistical data of experiments are linked to the more qualitative (albeit "empirical") content of an interpretative model of a theory (i.e some empirical reduct of the relevant interpretative model). There is, though, also a seeming correlation between Morgan's (ibid., p.208) remarks concerning the status of empirical models in econometrics and the status of empirical models in the natural sciences. She (ibid.) writes "... the empirical models that econometricians [work] with [are] a sort of halfway house, formed to capture the correspondence between theory and data ...". Well, empirical models in my terms also have a decisive role to play in "bridging" the gap between the interpretative models of a theory and the quantified data concerning some system in reality.

However, it is apparent that Morgan is indeed somehow conflating the notions of interpretative and empirical model set out above. She (ibid., p.201) sets out three points in the process of "finding an empirical model to match the theory involved [and] making the theory operational" (ibid.). Briefly they are transforming the "verbal theory" into mathematical form and deciding on values for variables, dealing with the *ceteris paribus* clauses under which the theory is taken to hold, and setting down the specific time frame for each theory. At least the first two points seem to be worthy of comment for our purposes. The translation of the "verbal" theory into mathematical form is in my terms done either at the linguistic level if Morgan means here the mathematical formalisation of sentences in natural language, or at the conceptual level, at the time of interpretation, if she means to emphasise the determination of values of variables. In either case this formalisation is not necessarily part of finding a suitable empirical model, although the specification of time frames as part of the specification of parameters, depending on the theory in question, might be.

A more serious difference is related to her point about the *ceteris paribus* clauses necessary for the theory to hold in the general way that it does. These conditions are already "cashed out" when an interpretative model of the theory is constructed in the sense that the "initial conditions" needed to make the specific calculations characteristic of actions regarding the interpretative model are formulated as part of the said model. The idealised character of such models is not so much a result of the *ceteris paribus* conditions at play at the theoretical level as simply a result of the fact that these models are still models of a particular theory. The empirical models in my terms are then even less related to such clauses. Of course, even when experiments are carried out, it is the case that certain factors are kept stable so as to be able to examine a specific aspect of the real system being studied. These factors are however far more specific than the *ceteris paribus* clauses and are rather a result of the specialising way in which science is typically practised. The relations between science and reality cannot be anything other than somehow idealised, given the complex nature of reality, although the construction of interpretative models of a theory starts the suspension of the generalisation implied by scientific theories' *ceteris paribus* clauses. More on this in Chapters 4 and 5.

A last remark concerning Morgan's article: she (ibid., p.207) remarks that Haalvelmo (1944) stated that "... in the absence of an experimental framework in economics, econometrics must act on both the theory and the data, making adjustments on both sides in order to get satisfactory models". In the natural sciences this is also true — even despite the presence of an experimental framework (or perhaps because of it?). The hierarchy of "tools" needed to bring this about is perhaps just of a different kind.

Let us now turn to the nature of the reality economics might or might not be about. Model-theoretically speaking science studies systems in reality. This refers to the abstracting simplifying nature of science. No-one, not even scientists (or economists), can study reality in all its fullness at once. Not only do scientists focus on some particular system of phenomena in reality at a given time, but also they aim to "adjust" that system in such a way that they can focus only on certain of its features. If certain abstractions are made from the richness of experiences that reality has to offer, scientific knowledge of the real system in question becomes possible. And, *vice versa*, if some knowledge claim is offered as part of science, the nature of that claim will (relative to the complexity of the universe) be simple and it will be about a sufficiently abstract version of some real system (even if that system is the cosmos!).

The problem with regard to the "social" reality of economics (versus reality in terms of "Nature" that the natural scientists focus on) lies in the "unstable" character of this reality as an "open", in the sense of being a so-called "uncontrolled", or "uncontrollable", system. The instability of social reality is often claimed to be the result of the multiplicity of various tendency or causal powers that are at play at the "uncontrolled" level of "open" "social" reality. Notice that although mainstream economists might be said to rely on the presumption that the economic system is stable, this is not stability at the level that I am talking of here. Also, this instability is not something that can be rectified by government intervention. Rather this is an inherent instability as far as social reality is concerned, i.e. an instability at the structural level of reality that is independent of our (either economists' or scientists') actions. And, anyway, even if I did agree with the above (which I don't) I do not think that economics, any more than science, should be about "stabilising" in the sense of somehow *changing* the complexity of *reality* into a controlled system. Rather it is about offering a glimpse as it were of some specific aspect of reality (as in the natural science case).

Can we ever test our hypotheses in the social sciences though? Backhouse (1997, pp.206,207) mentions four reasons why achieving a common understanding about reality (in terms of methodological studies in economics) is unattainable:

- Economists cannot "see" economic concepts in the same sense as mathematicians like Euler could "see" a polyhedron. This is definitely debatable, given the familiar debate about the status of theoretical terms, and the distinction between theoretical and observation terms in philosophy of science (see the next chapter).

- The continuously changing nature of economic reality (in Lawson's sense of social reality being dependent on human agency) makes it harder for a consensus to be reached. This is indeed (apparently) a problem in all the social sciences, and is what Bhaskar (1986) in the end identified as among the main sources of the limitations of naturalism.

- The need, as a result of the above, to work continually with assumptions that are unrealistic. Within a model-theoretic framework the "unrealisticness" of these assumptions and the idealised nature of their models become less of a threat to realism (see Chapter 5).

- Lack of agreement on the basic concepts that economics should be aiming to explain. This might be linked to the second point above, and definitely is more of a problem in the social than in the natural sciences. However, there is a question of degree at issue here — surely economists agree on certain basic phenomena even if these phenomena will be interpreted differently form different economic frameworks.[29]

Cartwright (in Davis, Hands & Mäki (1998)), when she describes social reality, claims that in the methodological sense the term "capacities" denotes abstract facts about economic factors in the sense of what these factors would produce if unrestrained. She (ibid.) gives the following example of a capacity in the economic sense: does progress in information technology have the capacity to increase income inequality by decreasing workers' bargaining power?, in other words, does it *tend* to increase income inequality even if that tendency may be offset by countervailing factors?

She (ibid.) identifies three characterising elements in the notion of capacities:

- potentiality: capacities describe what a factor can do in the abstract, not what happens in the real world;

- causality: capacity claims are not about co-association but about what results a factor can *produce*;

- stability: the ability to produce the effect in question must persist across some envisaged variation of circumstance.

She (ibid.) points out that statistical methods are not capable of providing necessary or sufficient conditions for any of the above, and that is supposedly why "hard" positivists tend to evade the issue of capacities. Cartwright however agrees with Mill that in economics this is not an issue that can be evaded. Economics should study causes "severally" (ibid.), since only then will it become possible to deduce results of any given (established) combination of causes or for any change in combinations. This implies the assumption that particular effects can be associated with each cause independently of its context and that when causes act in combination there is a rule that calculates their joint effects specific to each cause separately.

What kind of empirical support is there for capacity claims? Cartwright (ibid.) points out that even controlled experiments that isolate some specific factor under study can only make claims about the factor's behaviour in that specific experimental context. This is the problem of "transduction", i.e. the fact that the reasons for claiming that what holds under experimental conditions will hold elsewhere (or at other times), must come from *outside* the experimental context in question. In econometric terms this means that the direct use of a system of equations from data generated in one context to predict outcomes when the context is shifted clearly supposes that the estimated parameters describe capacities. Any statistical tests bearing on the stability of the parameters in question are thus relevant, although cross-contextual data are required. Cartwright (ibid., p.48) writes that

> Ideally, for purposes of empirical confirmation we should like the full scheme envisaged by Mill: knowledge of the capacities of all relevant factors present in some particular situation plus the rules of composition. Then we could deduce the expected behaviour and test our hypotheses by the conventional hypothetico-deductive method.

It is in this sense then that Cartwright (1989) claims that also scientific theories refer to causal capacities, rather than to sets of events if and only if the set of models — in which the theory in question is true, and which has some further class of substructures where imputations of causal capacities are concerned — is co-extensive with the set of models actually used as working interpretations of the theory in question. This relates to her notion of "socio-economic machines" (Cartwright, 1999, p.139). She (ibid.) writes:

> Models in economics do not usually begin from a set of fundamental regularities from which some further regularity to be explained can be deduced as a special case. Rather they are more appropriately represented as a design for a socio-economic machine which, if implemented, should give rise to the behaviour to be explained.

In the rest of this section I shall focus on the relations between mathematical and empirical models in economics, and the reality these models picture. Although it is true that few econometricians — or even economists — are at all positive about the actual impact of econometric results on the theorising of economists[30], econometrics remains, in a sense, the study of the "final" (quantitative) links between models and reality, and in the context of this article, something definitely then has to be said about the role of econometrics in "fitting" empirical models to reality.

The way econometrics predicts or explains[31] (i.e. talks about empirical links between models and reality) seems to imply, however, that many econometricians are interested solely in (abstract) econometric theory, and so not in applied economics. Whether econometrics then is the correct mechanism through which empirical links between economic models (theories) should best be addressed remains a debatable issue.

Econometrics indeed seems not to be about discovering quantitative laws concerning empirical regularities. It seems to be closer to (the theory of) mathematical statistics, than to empirical studies.[32] Interestingly, Keynes (in Moggridge, 1973), in correspondence with Roy Harrod, tells Harrod not to be reluctant to "soil his hands" (ibid., p.300) by examining the empirical relationships between models and empirical data. He (ibid.) writes

> The specialist in the manufacture of models will not be successful unless he [sic] is constantly correcting his [sic] judgement by intimate and messy acquaintance with the facts to which his [sic] model has to be applied.

Keynes however views the role of these "messy acquaintances" more in the context of evaluating the progress of economics than purely in terms of empirical motivations. In his discussion with Harrod regarding Tinbergen's models, he (ibid., p.299) claims, for instance, that the point of experiments in the natural sciences is to "fill in the actual values of the various quantities and factors appearing in an equation or a formula". He (ibid.) then points out that the situation in economics is rather different, and that "to convert a model into a quantitative formula is to destroy its usefulness as an instrument of thought". His motivation for this claim is based on the problem of replicating econometric findings. He (ibid.) writes:

> Tinbergen endeavours to work out the variable quantities in a particular case, or perhaps in the average of several particular cases, and he then suggests that the quantitative formula so obtained has general validity. Yet in fact, by filling in figures, which one can be quite sure will not apply next time, so far from increasing the value of his instrument [i.e. his model], he has destroyed it.

It might be that the complicated nature of the reality economic theories are "about" plays a role in upping the degree of difficulty of getting to content descriptions of model-theoretic links — more so than in the natural sciences — but surely that does not mean that *nothing* can be said about economic progress or success. Here, it seems, Keynes would agree. He (ibid., p.294) comments that

> ... suppose you have statistics covering a period of 20 years, what is required, it seems to me, is to divide these into convenient sections, say, of 5 years each, and calculate a proper equation for each period separately, and then consider what concordance appears between the different results. Until this has been done, a formula applying to the whole of the 20 years can have very little significance.

Again, it might be that the problematic nature of making claims regarding the predictive and explanatory abilities of economics should be seen against the fact that empirical data in economics are far more unstable than that of the natural sciences. This, however, still fits into a model-theoretic realism's picture of scientific progress. In such terms progress is measured in terms of various speeds at the different levels of science. Disciplinary matrices (or research programmes) change very slowly. Theories may change a little faster, interpretative models at an even higher speed and empirical models which are conceptualisations of empirical data, the fastest (see Chapter 5). Links between theories, models, and data can, however still be traced and articulated.

In the natural sciences experiments are about control, strict procedures, reproducibility, and also "replication". In this sense the verification or checking of experimental results becomes a fairly simple exercise. Any errors in calculation or carrying out experimental activities can usually be spotted and rectified with the minimum of fuss. Experiments in the economic sense of "replication" is a different story though. As Backhouse (1997, p.137) points out, where checking merely tests the experimenter's expertise and skills, replication tests the "existence of some phenomenon" (ibid.), in the Cartwrightian (1991b) sense that there is indeed some phenomenon to be explained. Replication naturally involves more than mere checking of experimental results then.

What is at issue is not the simple rerunning of some experiment, but, since replication only takes place as the result of some *contested* experimental outcome, and the correctness of the experimental procedure is in this case dependent on the outcome of the experiment, what is at issue is a repetition of the experiment in question with certain important differences. The replication will be identical to the established experiment in all of its aspects, *except* in certain "irrelevant" (ibid.) aspects regarding the identify of the experimenters, the location of the laboratory, and so on. Obviously this implies that what is termed "checking" in one context might be termed "replication" in a different context, depending on the issue at hand and the development of technical skills and apparatus.

The problem with all of the above in economic terms is of course that econometrics is on the whole non-experimental. Econometrics — in the sense of constructing estimators and analysing their properties — is mostly about data analysis. But, something else that must be taken into account in econometric terms is the differences in data. Economic data can either be experimental, cross-section data on individual markets (persons, households, or firms) derived form sample surveys, or aggregate time-series data. In the case of experimental data the situation would be much the same as in the natural sciences. In the case of cross-section data the complication that enters the situation is in the form of the econometric problem of regress. How does one know that one is, indeed, going "back" to the *same* world, i.e. to a world or reality that works in the same way that it did when previous experiments were carried out (previous surveys were done)? This problem is more complex in the case of aggregate time-series data, since here usually there is only one sample available — recreating issues that led to unemployment after the 1994 election in SA would for instance be rather difficult to do.

All of the above emphasises that econometric replication does not seem to have the ability to assure us that our representations (results of experimental analyses) do indeed represent reality. So-called "*a priori* economists", working in the positivist tradition, would say that in the nature of things this is not surprising, since in this sense empirical analysis is misguidedly seeking to concretise the metaphysical or theoretical. The only way that I can think of to address these issues is to analyse the relationships between theories, models, data (empirical models in my terms), and real systems far more vigorously than has been done up to now in economic methodological studies. The main problem that I see here is the fact that theorising and modelling seem to amount to much the same thing in economics. As mentioned before, economists often talk about theories and models as interchangeable and if they do reflect upon the relations between these stages of science, they find it especially problematic that activities of theorising and modelling are indeed as inseparable as they seem to be. Backhouse (1997, p.158) for instance remarks that "[e]conomists may defend theories, and it is theories which provide explanations, yet they analyse theoretical models and test empirical or econometric models". Also the non-monotonic analysis (offered in terms of a minimal model semantics in the next section) of different links between theories, interpretative models, empirical reducts, and empirical models as conceptualisations of data may resolve some of the above problems of economic replication.

Perhaps there are some overlaps between a model-theoretic analysis of (natural) science and philosophy of economics. The aim of science in these terms simply is depicted as offering certain idealised "insights" into the complex workings of "Nature". No statements about "absolute truth" or the unqualified "truth of scientific theories" are offered in such an approach. Rather, systems in reality may be explained in terms of certain models interpreting a certain scientific theory. Scientific theories cannot be universally true, but merely true in (a) particular conceptual (interpretative, mathematical) model(s) of it.

Moreover, these theories' statements are never meant to apply universally (or in a *ceteris paribus* way). Scientific theories are formulated *ceteris paribus*, however, their application and interpretation are context-specific, i.e. model-specific, and never "fixed" in any unqualified way. The closer we get to systems in reality, the fewer of these *ceteris paribus* conditions we need. Models are idealisations of real systems. However these ideal circumstances in which the theory is shown to be true are not as they are *by virtue* of the fact that all other things remain equal, but are rather the results of focussing on a specific real system and constructing a *particular* model driven by the scientific tradition, application goal, and other "thematic" factors present. Thus, the idealised *ceteris paribus* nature of theories and models are "relaxed" (Ross, 1999, p.254) by replacing (at least in all relevant respects for the application in question) these clauses with empirical models embedded into models interpreting the theories.

It is moreover precisely *because* of the idealised nature of models that we still have the possibility of having contact with reality given the abstracting way in which science and its enterprises operate. Idealisation does not mean universalisation, however. It is precisely because the model is so *specific* — in terms of its focus on selected features of some real system — that it is so ideal. This however in no way implies that no links with the "real complex" systems are possible, but rather that these links should be *established* and *checked* in a certain specific way, i.e. a model-theoretic way.

The slogan of a model-theoretic realism is "truth without universality". (See also Chapter 5.) This is meant in the sense that it is the specific model-theoretic kind of truth that is at issue, and that theories are never examined for their relevance to reality in their stark linguistic terms, but always in terms of their (conceptual) interpretations in their various models. Theories in this sense are not viewed merely as general knowledge propositions, but rather as the means of organising systems of their models in such a way that certain systems in reality can be (empirically) "embedded" into these models.

Perhaps interpreting the process of economics along these lines will solve some methodological issues, at least in terms of a realist context. I cannot, due to limitations of space, go into the realist-anti-realist debate in philosophy of economics here, except for noting some of the palyers in the field. There is quite a mixture of allegiances in philosophy of economics: From defenders of some kind of neo-positivist account of economics, such as (at various levels of extremes) Friedman ((1949), (1953)), Hutchison ((1960), (1977), (1992)), Rosenberg ((1985), (1986)), and Blaug ((1992), (1994)); to Hausman (1992) and Weintraub's ((1974), (1992)) accounts of economics; to realists (in various degrees) such as Lawson ((1989), (1994a), (1994b), (1997)), Mäki ((1992), (1993), (1994), (1996a), (1996b)), and Rappaport ((1988), (1993), (1996), (1998)); to McCloskey's ((1986), (1990), (1994)) rhetoric echo's of Fine's "natural ontological attitude".

All I can say here is that uncritical reductionism is the biggest enemy of any social science methodology and that a model-theoretic realism with its stage-by-stage fluid analysis of scientific enterprises, aided by its particular application of a non-monotonic logic minimal model semantics, offers considerable scope for the suppleness so characteristic of social science theorising. Whether or not economists should be more than observers or data-collectors, I cannot say. Should one answer "yes", to the above question (i.e. should one think that economists must be more than observers or data-collectors), then economists have the task of explaining economic phenomena, which would bring them into the realist debate. And in this sense, because model-theoretically the "non-accurate" nature (in the sense of exaggerating certain features, or minimising others) of interpretative models precludes a one-to-one relation of correspondence between real systems and these models, a model-theoretic interpretation of the role of models might indeed prove to be a handy methodological device in economic methodological studies.[33]

I don't know whether economic models are more successful than not in incorporating empirical substructures that enable economists to make predictions and explain economic phenomena. However, neither the difficulty of isolating economic phenomena, nor the fact that (perhaps more frequently in economics than in physics) a given model might work magnificently in one experimental setting (see Ross, 1999, p.255), but not at all in another even if both experimental situations lie within the domain of the theory in question, precludes a realist interpretation of economic theory.

Cartwright (1999, p.157) writes:

> Economists simply do not know enough to fill in their law claims sufficiently. ... Laws in the conventional regularity sense are secondary in economics. They must be constructed, and the knowledge that aids in this construction is not itself again a report of some actual or possible regularities. It is rather knowledge about the capacities of institutions and individuals and what these capacities can do if assembled and regulated in appropriate ways.

I do not deny that (metaphysical) notions such as capacities or tendencies in Cartwright, Bhaskar, or even Tony Lawson's senses might be needed in philosophy of economics. Whether they aid realist quests, is however another question. Be that as it may, I claim that (with or without these notions) at least reconstructing the different stages of the process of economic modelling in terms of a model-theoretic realist interpretation of science might clarify methodological factors at issue in a realist context, and might even make talk of realism in economics seem less unattainable.

Cartwright (ibid., pp.3,4) remarks that

> ... theories in physics and economics get into similar situations by adopting opposite

strategies. In both cases we can derive consequences rigorously only in highly stylised models. But in the case of physics that is because we are working with abstract concepts that have considerable deductive power but whose application is limited by the range of concrete models that tie its abstract concepts to the world. In economics, by contrast, the concepts have a wide range of application but we can get deductive results only by locating them in special models.

Again, the features of economic methodology discussed above as well as the importance of the interplay between exogenous and endogenous factors in economic models, perhaps may all be considered more fruitfully via an application of non-monotonic logic in terms of a minimal model semantics within a model-theoretic context (discussed in the next section).

7. THE PROBLEM OF "OVER-DETERMINATION" OF THEORIES BY EMPIRICAL MODELS[34]

Almost all projects aimed at demarcating the "purely" observational (in the sense of so-called "raw sense data") from the theoretical are beset by certain difficulties which are invariably the result of two major issues. On the one hand, these difficulties arise as a result of the nature of the links postulated to exist between these two kinds of entity and the languages they are described with, and, on the other hand, the difficulties are caused by the nature of the set of so-called "intended applications" of a theory, especially in terms of the existence of more than one so-called "empirical model" as the "real" domain of reference of the terms of theories. I shall show in this section that a model-theoretic realist analysis of the structure of scientific theories may clarify the motivations behind choices for certain empirical models (and not for others) in the above context of demarcation in a way that shows that "disentangling" theoretical and observation terms is more deeply model-specific than theory-specific. (See also Chapter 3.) A mechanism to trace "empirical choices" and their particularised observational-theoretical entanglements will be offered in the form of Shoham's version of non-monotonic logic.

As mentioned above, a model-theoretic realist account (see also Ruttkamp (1999)) of science places linguistic systems and their corresponding non-linguistic structures at different stages or different levels of abstractness of the scientific process. Instead of looking towards typical statement approach's notions of correspondence rules or bridge principles to address observational-theoretical translations or referential questions concerning terms in theories, a model-theoretic approach acknowledges the re-interpretability of the language(s) in which theories are formulated and so turns towards mathematical models of theories as the crucial links in the interpretative and referential chain of science (see Chapter 3).

Merely "presenting" the theory "in terms of" its mathematical structures (or the set-theoretical predicates representing the class of these structures) typical of the so-called non-statement accounts of theories is not considered sufficient, since these accounts seem to eliminate — or at least de-prioritise — the possibility of addressing within a realist context the nature and role of general terms and laws — expressed in some appropriate formal language — in science. Model-theoretically speaking, this is unacceptable, since the links between the terms of scientific theories (as linguistic entities) and their interpretations in the various models of these theories in this context are taken to regulate the whole referential process, since such links offer particularised theoretical/ observation distinctions.

Obviously, from previous sections, the reconstruction of the experimental (empirical) stage of science is rather more problematic in comparison to the theoretical stage which may be axiomatised "quite easily" with the help of either set-theoretic predicates or structuralist theory-cores. Moreover, as Nagel (1961, p.98) writes:

> When a theory is formulated by way of a model, the language used in stating the model usually has connotations that the language of experimental procedure does not possess. Thus, ... the expression in the Bohr theory referring to electron transitions is not equivalent in meaning to the expression referring to spectral lines. In such cases, accordingly, since the defining and the defined expressions in explicit definitions are equivalent in meaning, it is most unlikely that rules of correspondence can provide such definitions.

And, as we know, there is the rub. Not only is it necessary to find a way to move from intensional linguistic definitions of terms in theories to extensional definitions of these terms, but also there is the problem of linking the language of science to the language of experimental (empirical) procedures.

Inherent to a model-theoretic account of science, is the notion of mathematical "reducts" (see Kuipers 2001, p.345), i.e. mathematical structures created (in model-theoretic terms) by leaving out in the language and its interpretations some of the relations and functions originally contained in these entities (see Section 2.5) — best exemplified perhaps by the structuralist notion of partial potential models (see Chapter 4) — which makes sense of worries concerning connotations in the "language used in stating the model" not present in the "language of experimental procedure". Furthermore a model-theoretic realism continues on to link the empirical language of experiments with the theoretical language of science by its notion of an isomorphic relation from empirical models (conceptualisations of data) into certain substructures (model-theoretically so-called "empirical reducts") of the interpretative models of theories.

In this section I want to show how a model-theoretic account of scientific theories, augmented, at the level of empirical reducts, by the machinery of non-monotonic logic as developed in particular by Shoham, may enable us to express reference relations between theories and empirical (observational) models in the face of theory change in general, and multiple model choice in particular. Although reflecting on issues concerning both the progress and — perhaps, more importantly — the process of science seems often — especially in formal contexts — to end somehow in deciding on issues of truth approximation and theory reduction, I shall not dwell on these issues here. Rather than focussing on progress in terms of a gradings of truth and success, I want to focus on the choices made when faced with more than one empirical model and the motivations for these choices.

Hence, I shall concentrate here on the question of temporary information (or perhaps temporary knowledge) in the presence of empirically equivalent empirical reducts and empirical models and so, in the final instance, in model-theoretic terms, of data "over" determining theories. Against the background of the formal fact that there are many more than just one model offering a true interpretation of one given theory, I shall focus especially on the empirical proliferation of certain empirical parts of models, in the sense that the empirical adequacy of a theory may not manifest itself uniquely.

My answer when confronted with questions concerning model choice has usually been that these are about very particular concerns that will depend on the particular intentions of a particular scientific community at a particular time — notice the echoes of the structuralist concerns (Chapter 4) regarding the limits of the mechanisms of pure semantics to present these intentional choices. Although I still claim this to be the case, I have always been dissatisfied with the — at least apparent — informal character of such an answer. In this context, I want to consider with you the possibility of introducing into the empirical equivalence debate the non-monotonic mechanism of default reasoning, refined into a model-theoretic non-monotonic logic (based on the logic of Yoav Shoham) offering a formal method to rank models.

In terms of what I call "temporary knowledge" we need at least to consider the following questions: Where in the process of science would we find these particular pockets of temporary knowledge? In what sense exactly may scientific knowledge be temporary? How does such knowledge affect our final judgements on the nature of scientific progress?

Briefly, in answer to these questions: Where do we find such pockets of temporary knowledge? We find such knowledge everywhere in the process of science, obviously, since we know that even the "best" theory at a certain time might in all probability be refuted at some point in the future. However we find the most extreme form of it at the level of the process of science where empirical adequacy is determined, that is, in my terms, the level at which we are considering so-called "empirical reducts" and their relations to so-called "empirical models".

The sense in which I mean this knowledge to be "temporary" is the one in which we make choices for certain models (and so sometimes for certain theories) at certain times. The context for this discussion is that of empirical equivalence in Van Fraassen's sense of the notion: He (1980, p.67) writes: "If for every model M of [theory] T there is a model M′ of T′ such that all empirical substructures of M are isomorphic to empirical substructures of M′, then T is *empirically at least as strong as* T′ [*sic*]" — put in this way it seems rather as if it is T′ that is empirically at least as strong as T. Earlier Van Fraassen (1976, p.631) wrote that "Theories T and T′ [each being as least as strong as the other in the above sense] are *empirically equivalent* exactly if neither is empirically stronger than the other. In that case ... each is empirically adequate if and only if the other is".

But what is the status of the models or empirical reducts — or even the relations of empirical adequacy — we do not choose at a specific time then? The knowledge or information about the particular empirical model(s) in question that they carry, certainly *is* still knowledge, is it not? Well, yes and no. Fact is we need a formal mechanism by which we can depict our choices, the motivations for our choices, and the change of both of these, should the context within which we are applying some theory, change.

This is the sense in which I speak of knowledge as temporary — we choose to work with a certain model or empirical reduct at a certain time, but we may always change our minds and make a different choice which might imply a change in the set of knowledge claims (and the meta-tracings of reference links and theory-observation distinctions) our theory is offering, and this is where non-monotonic logic in the form of default reasoning comes in, as I shall explain below.

Related to this, recall that as far as the nature of scientific progress is concerned, my (multi-level) view is the following. Theories change very slowly, interpretative models more quickly, and empirical reducts and the empirical data bases (the accumulation of empirical data via observations and experiments) they depict, the quickest. (See Chapter 5.) In this sense, I agree with Kuhn that neither the content of science nor any system in reality should be claimed to be "uniquely exemplified" by scientific theories from the viewpoint of studies of "finished scientific achievements". And, therefore, one has to accept the open-endedness (see Chapter 3) of theories as a permanent feature of the total process of science. Notice though that this open-endedness to me is represented by the ebb and flow of the models (including their empirical reducts) of the theory which ensures the continuity of science at least at a formal (meta) level of analysis.

Hence I imply that issues of theory succession or reduction are often for long periods of time better — or at a finer level of analysis — interpreted as issues of model succession or reduction, and that this implies that certain aspects of our knowledge are more temporary than others. I claim the terms of an already established theory can be said to be "about" an ongoing potential of entities in some system of reality to give reference to some objects and relations in *any* model of that theory. The actualisation of this potential requires human action in the sense of finding and finally articulating "satisfying" referential relations between systems in reality and certain empirical aspects (reducts) of models of the theory (see Chapter 5). And it is the nature of these referential relations that will be the topic of the rest of this section.

Let us now focus on what I term "empirical proliferation". In a sense this is the reverse of the traditional under-determination of theories by data scenario. In philosophy of science the issue of the under-determination of theories by data is the original problem of explaining — and perhaps justifying — the existence of empirically equivalent, yet incompatible, scientific theories. In the history of science instances of such theories are quite common — think of the various ways in which an electromagnetic field has been described from Faraday through Einstein to Feynman.[35] The bottom line is (in this context) that empirical data are too incomplete to determine uniquely any one theory.

Now also keep in mind that contact between scientists and real systems which result in scientific data is relative to the state of scientific knowledge and of technological development at the time, as well as the research tradition or disciplinary matrix in which scientists work at that given time. In other words, scientific knowledge is amendable and even defeasible, because of its contingent and particularised links with the reality it describes (and explains). Recall that according to Van Fraassen (1976, p.631) a theory is empirically adequate if "all appearances are isomorphic to empirical substructures in at least one of its models". This view leads the way for the model-theoretic interpretation of empirical equivalence according to which theories with the same empirical reducts or at least some empirical models, are empirically equivalent.

These definitions point to the reverse case of traditional under-determination of theories by data, i.e. a specifically model-theoretic interpretation there-of — i.e. under-determination of data by theories. This view focusses on a slightly different aspect of traditional empirical equivalence. In general, as we have seen, scientific theories, depicted as syntactic (linguistic) entities that need to be interpreted to be given semantic meaning and reference, are not able to capture uniquely their semantic content. Within a model-theoretic context two kinds of interpretative relation come into play whenever questions of theory application have to be addressed. The first set of interpretative relations exist between the terms in some theory and their extensions in its various models. These relations assign meaning (and potential reference) to the theoretical terms. The second set of relations exists between the terms of models (or of only one model), via an empirical reduct of that (those) model(s), and the objects and relations of some real system (or systems). Hence, model-theoretically speaking, theories are over-determined by data and by their individual models.

Retaining the notion of scientific theories as linguistic expressions at the "top" level of science solves the problems regarding the justification of the existence of many (conceptual) models as interpretations of any one theory by the simple (formal) fact of the incompleteness of formal languages. Thus the possibility of a given scientific theory being interpreted in more than one mathematical model (structure) is natural in a very basic sense in model-theoretic terms. The second proliferation of relations between models and their empirical reducts and between these and empirical models may also turn out to be less counter intuitive than might be thought at first glance, if it is understood that the possibility of articulating a chain of reference is *not* jeopardised under such circumstances.

Recall now that in model-theoretic realist terms theories are empirically adequate if and only if they are true in certain models some of the empirical reducts of which conceptually encompass the empirical data of the relevant real system. In this sense the first step of the model-theoretic way to confront the model-theoretic over-determination implied by either the choice of a model for interpreting a particular theory, or the choice of a model into which certain empirical data are embedded, is to keep the following structural fact regarding the scientific process in mind. The choice of empirical reduct has to be such that it has embedded into it (an isomorphic copy of) some empirical model in which certain "observation" sentences are true. However, simultaneously, the mathematical model of which this empirical reduct is a substructure must be one that "makes" or "keeps" the sentences in the language of the theory that is shown to be empirically adequate, true as well.

This two-way characteristic of a model-theoretic analysis of scientific realism ensures that tracing theory-model-reality links — even if presenting a rather complicated undertaking — is still articulable (see *Figure 1* in Section 2.5). Simultaneously this shows however also the complexity of theory-model-data links. In what follows I claim in particular that an application of non-monotonic default logic to situations of over-determination of theories by models and data may enable us to formalise and get a grip on this complexity. I shall show that this application entails applying a particular kind of default rule — by defining a so-called total pre-order relation on the relevant set of possible worlds of the relevant language (see the definitions in Section 2.2) — at the level of empirical models which I claim results in a particular kind of ordering or ranking of these models in terms of preference determined by the content of this rule. My claim shall further be that this ordering induces an ordering both of empirical reducts and models of theories themselves, and may ultimately even result in a ranking of theories.

The context of looking to non-monotonic reasoning as a possibility of rationalising model choice is that of abduction.[36] Simply put, in the face of over-determination of theories by empirically equivalent models, we are faced with a situation analogous to inference to the best explanation, since we have a "theory" but have to choose under certain particular contingent circumstances, out of many options one empirical reduct — and first model — via which it (i.e. the theory) is linked to a particular empirical model and so to a particular system in reality. Kuipers (1999, p.307) states that abduction is "the search for an acceptable explanatory hypothesis for a surprising or anomalous (individual or general) observational fact". The fact that our knowledge at the level of empirical models is finite and incomplete and therefore changeable does *not* however imply that we cannot discover *some* rational aspects of the kind of abductive reasoning required in this context.

Yoav Shoham (1988, p.80) points out that in certain issues regarding incomplete information, we should concentrate on distinguishing between the meaning of sentences on the one hand, and, on the other, our reasons for adopting that particular meaning and no other. The latter will naturally be outside the domain of the system of logic we are working in at the time. I agree and acknowledge the contingency of the factors determining the nature — and choice — of a certain model at a certain time. But in my terms, this is a matter, though, to be articulated or pinpointed via the empirical models of the theory about the construction of which admittedly not much can be said external to some particular context of application of the theory in question. Once confronted with more than one empirical model though, I claim we may make use of Shoham's kind of extra-logical motivations to rank these empirical models in a certain order.

Formalising this a rather complex task. One way in which to do so might be to take all existing possibilities present at a certain time into account, and summarising the reasons for picking a certain empirical model — and so a particular empirical reduct of a certain model — at a certain time in such a way that the existence of other models — and other empirical reducts — is not denied, but simply, for a certain period of time, put on hold as it were. A method for doing this is offered to us by the nature of non-monotonic logic in general. In particular for our purposes here Shoham's model-theoretic non-monotonic logic is preferable, since it offers a fairly simple way of ranking models, which perhaps is not as adequately possible in other versions of non-monotonic logic.[37] A non-monotonic logic consists (for our purposes) of a propositional language over a finite set A of atoms, together with a minimal model semantics. This semantics allocates truth values to sentences with the aid of the usual valuations, but uses a total pre-order on the valuations to define a new semantic consequence relation between sentences, namely the defeasible entailment relation. (See the formal definitions given in Section 2.2.)

The general idea behind Shoham's reasoning that I find has some appeal in our context is (ibid.) that sometimes it is necessary to take "decisions" in our reasoning, while ignoring some information that is potentially relevant, but at the same time accepting or expecting to "pay the price of having to retract some of the conclusions in the face of contradicting evidence" (ibid.). The trick is to have some rational way of keeping track of these retractions. Gerhard Schurz (1995, p.285) writes:

> Applied laws or theories make claims of the following sort: (i) in the normal case, for all applications of the kind A(x) the theoretical claim r(x) or the predicative claim P(x) will be satisfied, but (ii) there are exceptional cases, although (iii) it is neither possible nor sensible (on complexity grounds) to give a complete classification of them or to list their probabilities. Hence, all that one can do is to understand the loose if-then relation ... as an uncertain but qualitative, i.e. non-probabilistic, implication of the form ... Normally, if A(x), then B(x), formally: $A(x) \Rightarrow B(x)$.

Traditionally, logic is concerned with cautious and conservative reasoning. It finds its natural home in mathematics, the theorems of which are immune to fashion and the passage of time. But life in general and science in particular need more than mathematics — we need common sense and contextualisation. This involves the capacity to cope with situations in which one lacks sufficient information for one's decisions to be logically determined, so that one has to try to distinguish between possibilities that are more plausible (i.e. "normal") and those that are less plausible at a given time.

Shoham (1988, pp.71-72) sets out his non-monotonic scheme as follows:

> The meaning of a formula in classical logic is the set of interpretations that satisfy it, or its set of *models*[38] .... One gets a non-monotonic logic by changing the rules of the game, and accepting only a subset of those models, those that are 'preferable' in a certain respect (these preferred models are sometimes called 'minimal models' ...). The reason this transition makes the logic non-monotonic is as follows. In classical logic $A \models C$ if C is true in all the models of A. Since all the models of $A \wedge B$ are also models of A, it follows that $A \wedge B \models C$, and hence that the logic is monotonic. In the new scheme we have that $A \models C$ if C is true in all *preferred* models of A, but $A \wedge B$ may have preferred models that are not preferred models of A. In fact, the class of preferred models of $A \wedge B$ and the class of preferred models of A may be completely disjoint! Many different preference criteria are possible, all resulting in different non-monotonic logics. The trick is to identify the preference criterion that is appropriate for a given purpose.

The process of making informed "guesses" or "choices" on the basis of a mixture of definite knowledge and default rules is called defeasible reasoning. The word "defeasible" reflects the fact that our knowledge may be amended, in other words that a certain default rule may be "defeated" by exceptional circumstances, or a change of circumstances caused by a change in the content of our knowledge. Defeasible inferences are inherently non-monotonic, since amending our system of knowledge might change our conclusions. For example, Shoham (1988, p.25) writes:

... from the fact that a ball is rolling in a certain way we infer that it will continue to do so, but if we add the fact that there is another ball directly in its path we change our prediction. This is why defeasible inferences cannot be represented in the classical logics.

As an example of the need to go beyond the irrefutable logical consequences of one's definite information, consider a simple physical light-fan system[39]. Say we take an ordinary two-valued propositional language with atoms p and q, where p: the light is on, and q: the fan is on. p can be T/F (1/0) or q can be T/F (1/0) such that the four possible states of the system are depicted by the set **W** = {11,10,01,00} (where a specific valuation depicts a specific state of a system). Say now, we determine theoretically that it is the case that p $\vee$ q, this reduces the frame of our language to {11, 10, 01}. Then we — or some of us at least — discover say, in reality, that we can see whether the light is on, but are too far away to see or hear whether the fan is on. Thus we have limited knowledge about the system. Now suppose the system is really in state 11, i.e. that the light and the fan are both on. We will know only that the light is on, i.e. that p is the case, not that both components are on, i.e. not that p and q are both the case. Our definite knowledge suffices to cut our current frame of states down even more to the frame consisting of the models of p, i.e. Mod(p) = {11, 10}. So far, so good. Where's the problem?

Suppose we urgently need to know what the state of the system is, because state 10 is a state requiring that the agent take urgent measures of some sort. This implies that we want to cut down the frame Mod(p) = {11, 10} to a frame with just one element in it. We need to go beyond our definite (although incomplete) knowledge, but without making blind guesses. How can we do this in a reasoned way? We can use a default rule such as "Experience and descriptions of the system have shown that when the light is on, the fan is normally on also" to make the informed guess that the state is actually 11.

Exactly how do default rules justify cutting down the set of models of our definite knowledge though? Or rather, what would we be willing to regard as a default rule? After all, not every rule of thumb can be taken seriously as a default rule. Shoham (1988) requires that a default rule should be expressible as an ordering on possible worlds (or models). (In the context of our example, the possible worlds are just the states of the system, namely **W** = {11, 10, 01, 00}.)[40] Shoham focusses on using non-numerical default rules, such as the rule "11 is more normal than 10, which in turn is more normal than 01 and 00" as the basis for "informed guesswork". All we require is that the rule arranges the states of the system in levels, with the most normal states occupying the lowest level, then the next most normal states, and so on, until the least normal, least typical, least likely states are put into the top level. The given rule yields the ordering:

$$01 \quad 00$$
$$10$$
$$11$$

Now we can choose between the two models of p in our previous example, because 11 is below 10. Our choice reflects not merely our definite knowledge that p is the case, but also our default knowledge that 11 is a more preferred state of the system than 10 is (by the default rule stated above).[41] (See Section 2.2 for formal definitions.)

In summary, default rules may be used to justify defeasible reasoning as follows: Order the possible states of the system from bottom to top in levels representing decreasing preference; given definite knowledge α, look at the states in Mod(α) — the set of all models of α; pick out the states in Mod(α) that are *minimal*, i.e. lowest in the ordering; then any sentence true in each of these minimal models of α may be regarded as plausible, i.e. as a good guess. So, whereas α classically entails β, i.e. $\alpha \models \beta$, when among ALL the models of α no counterexample to β can be found, α defeasibly entails β when among all the most PREFERRED models of α no counterexample to β can be found.

Note though that a default rule is not an absolute guarantee. Our informed guess may turn out to be wrong. Normally if Tweety is a bird then Tweety is able to fly. But exceptional circumstances may defeat the default rule. Tweety may be a penguin or an ostrich. Tweety may be in Sylvester's tummy. Abnormal states or a change in the content of the body of knowledge concerning a certain situation can sometimes occur. That is why, after all, in such cases we call our reasoning "defeasible".

Now, back to the context of science, given all of the above, the possibility of after the fact semantic reconstructions of reference links from theories to some real systems formulated with the help of model theory and non-monotonic logic (in terms of a minimal model semantics) offers a way to get us out of at least some of the apparent difficulties implied by over-determination and empirical equivalence in the model-theoretic way, as follows. In the scientific context I claim a default rule containing at least the following two conditions — or orderings — might be useful.

The first condition induces an ordering or ranking of empirical models in terms of precision or accuracy. This condition has to do with the highest quality of data and the finest level of technology. For now, I am considering cases here where we have to choose among different equivalent empirical models of which all may be embedded into the same reduct, or at least empirical reducts of the same type. The second condition that I would include in my default rule, is more often concerned, together with a choice of empirical model, also with a choice of empirical reduct, since here the condition implies a ranking of empirical models that may induce a ranking of empirical reducts. Here the rule states that empirical models that can be embedded into empirical reducts of a type that contains a larger class of empirical terms from the theory than others, are preferable.

The second condition has two noteworthy implications. First it shows how such a ranking distinguishes between weaker and stronger links between theories and reality, since a theory that is model-theoretically linked to an empirical model embedded into an empirical reduct containing a larger class of empirical terms than others, may be said to be more effectively "about" some real system than would otherwise be the case. Also, in terms of the progress of science it might be preferable to have a mechanism justifying including into a particular model of a theory previously exogenous factors as endogenous ones. (Think of the problems related to such changes in philosophy of economics, and how a non-monotonic preferential analysis at that level in that context might impact on resolving those problems.) This becomes a possibility if we enlarge the type of empirical reducts.

Placing both these conditions together into one default rule we may find that the resulting rankings of empirical models induce rankings of empirical reducts, which might induce rankings of models themselves, and which, may, ultimately, induce rankings of theories. Let us look at a simple example, again in terms of our light-fan system.

Theory: $p \vee q \equiv T$

- Empirical situation: Only the light can be observed

    This implies that

    - p: empirical term
    - q: theoretical term

Models of T	Empirical Reducts	Empirical models
11	1-	1-
10	1-	
01	0-	

- The observation of the light in an on position, cancels the empirical reduct 0-, which in turn cancels the model 01
- Our choice of empirical model thus induces the following ordering of empirical reducts:

    0-
    1-

    and the following ordering of models:

    11 10

- This changes our theory to $T' \equiv p$
- Say the empirical situation is enhanced by developments in technology and we can observe that whenever the light is on the fan is off. Then our frames of models become

Models of T'	Empirical Reducts	Empirical Models
11	11	10
10	10	

- The result of our observations now is that the empirical model 10 "cancels" the empirical reduct 11, and this, in turn "cancels" the model 11
- Our new enhanced empirical model now induces the following ordering of

empirical reducts:

> 11
> 10

and the following ordering of models:

> 11
> 10

- This changes our theory to $T'' \equiv p \wedge \neg q$

Another example: In the evolution of science from Newtonian mechanics to Einstein's general theory of relativity it became obvious at a certain point that no empirical reduct of any Newtonian model could accommodate certain empirical models which could be accommodated in empirical reducts of the general theory of relativity. This led, in the end, to Newtonian mechanics making way for — or perhaps, "evolving" into — the general theory of relativity. Notice nevertheless that the status of certain Newtonian models as approximate models of Einstein's theory of relativity and the approximate embedding of Newtonian empirical models into certain Einsteinian models remain untouched. The mechanism of non-monotonic logic according to which the state of — momentarily, or contextually — less preferred models as "models" of a theory still is acknowledged, thus offers a new way of reflecting on the accumulation of scientific knowledge.

Recall that on my view of scientific progress generally theories change much slower than models. Specifically, theory changes usually occur only when the possibility of changing and modifying the models of the theory concerned has been exhausted, which confirms the continuity of scientific knowledge. Think of the rotation of the orbit of Mercury. Einstein's theory of general relativity interprets all gravitational interaction as due to the curvature of space-time. Applying Riemann's mathematical theory of curved spaces (of any number of dimensions) to the physically real curved space-time — a four-dimensional space-time — and correlating by equations the so-called "curvature tensor"[42] of the space-time continuum with the distribution and motion of masses, all of Newton's law of gravity's results may be derived at the first approximation. For example, according to Newton's law of gravity planets move along the elliptical orbits with the sun in one focus that Kepler's empirical laws had already determined.

According to Einstein's theory of general relativity though all motions should be considered in the four-dimensional world of "events" (x,y,z,ict)[43] which is curved if gravitational fields are present. The "world lines" Gamow (1962, p.205) of any material body in the four-dimensional world (representing the history of the motion of that body) must be geodesics or the "shortest lines", and can be calculated on the basis of the general theory of relativity. Exact calculations showed that the elliptical orbit of a planet around the sun does not stay stationary as Newton predicted. Rather it is "slowly rotating with its major axis turning by a small angle in the course of each revolution" (ibid.). This phenomenon is most noticeable in the case of Mercury, which is closest to the sun and has the most elongated orbit of all the planets. In other words, no model of Newton's laws of motion and his law of gravitation could solve the anomaly with regard to Mercury's perihelion motion, and so finally a new theory was formulated, some of the models of which can indeed explain these discrepancies. Gamow (1962, pp.205-206) writes

> Einstein calculated that the orbit of Mercury must turn by 43 angular seconds per century, and solved herewith the old riddle of celestial mechanics. It was calculated by mathematical astronomers long before Einstein was born that the major axis of Mercury's orbit must slowly turn around because of the perturbations, i.e., gravitational disturbances, of the other planets of the solar system. But, there was a discrepancy between the calculations and the observations amounting to 43 angular seconds per century which could not possibly be explained [in the models of classical mechanics].

Returning to my conclusions from the above, I claim that non-monotonic preferential default rules and consequent rankings enable us to reduce both the available — or possible — choices of models, empirical reducts, and empirical models. This kind of analysis offers a method of getting an articulable grip on empirical equivalence of any kind. The minimal model semantics of non-monotonic logic fulfils what Kuipers (1999, p.307) calls the "main abduction task", i.e. "the instrumentalist task of theory revision aiming at an empirically more successful theory, relative to the available data, but not necessarily compatible with them" (ibid.), although this is done here mostly through revision — or change — of relations of empirical adequacy implying possible revision of choices concerning empirical models, empirical reducts and (conceptual) models.

Although the above application of non-monotonic logic starts at a finer level of analysis than is usually the case in non-monotonic contexts (where we simply look at rankings of the states — models — of the system in question), the model-theoretic structuring of relations between models, empirical reducts, and empirical models makes possible the kind of "carrying over" of rankings that I have set out above.

Notice that relations of empirical adequacy is thus temporary and contextual, as Laudan and Leplin (1991) also concluded in their 1991 article entitled *Empirical equivalence and under-determination*. Science progresses fastest at the level of empirical models, but continuity is ensured by the fact that these models remain conceptualisations of observations, even if these observations are also contextual. The point of a model-theoretic realism is exactly that instead of offering simply one intended model of "reality", a theory is depicted as a way of constructing or specifying a collection of alternative models, each of which may represent, explain, and predict different aspects of the same (or different) real system(s) via the same or different empirical reducts isomorphically linked to the same or different empirical models.

In the above we have mostly concentrated on cases of empirical equivalence in terms of model-theoretic over-determination. What — in terms of realist concerns — about under-determination in the traditional (Laudan/Leplin) sense? (I.e. — different theories, same empirical model.) Applying non-monotonic logic within a model-theoretic context also helps to minimise traditional under-determination of theories by models and data within a context of scientific progress, since it leads to choices of more accurate, more encompassing (empirical models and so) empirical reducts, and in certain cases, it may even help to eliminate certain models entirely.

In this sense — in a realist context — a scientist can "know" — or at least determine — that she is working with the "same phenomenon", even if using "different" theories or "different" models, because of the possibility of analyses that a model-theoretic minimal model semantic realism offers of the different empirical links between different empirical models of different (conceptual) models of (perhaps) different theories. Detailed analyses of these empirical links will reveal common factors on the reality side of the link (e.g. light blobs observed through different telescopes by different people at different times indicating — by careful analyses — a common factor called "Neptune") which entails the "same phenomenon". And, moreover, cases where the same empirical model is embedded into different empirical reducts also show the continuity of science at the empirical level. Kepler took Brahe's precise empirical observations, i.e. the empirical data forming the empirical model of the theories in terms of celestial spheres that Brahe worked with, and fitted these data — i.e. Brahe's empirical model — into his (Kepler's) theory in terms of elliptical orbits.

We can thus even in the face of the fact that our fallible sensory experience and the finiteness of experimental data at a given time indicate that our knowledge of reality at such a time is limited, contextual, and temporary, rationally discuss the choices we make concerning so-called "empirically equivalent" models and keep track of changing theory-observation distinctions. It might be then possible, after all — contrary to Popper — to give some kind of rational motivation for the so-called "creative" leap that we make from data to theories. Of course a preferential analysis of empirical model choice admittedly does not "simulate" the "processes in the minds or brains of scientists" (Kuipers, 2001), but it does make sense of the motivations for certain of these scientists's actions based on the status and development of the knowledge claims they make.

Niiniluoto (1999, p.4) writes:

> ... for the most part, scientific activities do not involve belief in the sense of holding-to-be-true: rather ... scientists propose hypotheses and pursue research programmes in investigating the limits of the correctness of their theories. However, such enquiry is always based upon some assumptions of 'background knowledge'. If successful, it will also have tentative results, in principle always open to further challenge by later investigations, which constitute what is usually referred to as the 'scientific knowledge' of the day.

While he offers a theory of truthlikeness to depict this (above) account of science, I depict an account that is basically the same, but based on preferential model-theoretic realism.

In general, I conclude then that scientific theories may indeed say something about reality, but it is not possible when faced with an *uninterpreted* theory and possibilities of over-determination of the theory by both data and models to determine or claim that it will definitely or uniquely be applicable to a certain aspect of reality and to no other. The model-theoretic notion of articulated reference and truth augmented by non-monotonic preferential mechanisms to get a grip on empirical over-determination, may render the process of science expressible to rather finer and more accessible detail than may be possible on other accounts of science. When reference is traced via model-theoretic relations between theories, models, and data, and extra-logical default rules are used to formally order our choices in a rational responsible way, Quine's inscrutability of reference becomes an even vaguer notion than before. Hence reference — at least in this sense — does not appear to be indeterminate after all. Secondly this implies that the content of the meta-verification procedures for the processes of science cannot be given uniquely, but is rather a result of the context-specific actions and constructions of human scientists. In other words, theory-observation distinctions — or the definition of c-rules — remain somewhat less precise than one might wish for in a positivist sense, but overall at least these distinctions remain articulable in the model-theoretic sense — which is more important for the success of a realist quest. (See Chapter 3.)

It might be that a model-theoretic realism aided by a minimal model semantic (non-monotonic) ranking of models (empirical reducts and empirical models) offers, at least partly, some response to Laudan and Leplin's (1991) concerns about the "collapse" of epistemology into semantics in terms of traditional under-determination and empirical equivalence issues, taken almost as two sides of the same coin. Non-monotonic default rules are extra-logical and are determined by the state of knowledge of a system at a particular time (i.e. "the agent knows that the light is on"). The new perspective on the consequence (entailment) relation that non-monotonic offers might thus present us with a different way of looking at Laudan and Leplin's (1991) claim that evidential support for a theory should not be identified with the empirical consequences of the theory.

The following remark (Stegmüller 1979, p.126) sums up the above very adequately:

> It is theoretical reason which supplies detailed work and which also recognises a new theory as progressive in respect to an earlier [one]. But it is practical reason which decides which of the different possible ways promising progress is to be taken. In my [Stegmüller's] opinion Kuhn is completely right when he emphasises [by his thesis of the primacy of practical reason] the role of value judgements in the development of science. Where science develops and possibilities bifurcate, no mere brown study will suffice. Rational decision must enter in. That a new theoretical beginning will actually bring progress, can, in time, only be believed and hoped.

Against all of the above, and to clarify the distinctions and similarities between a model-theoretic realist account of scientific theories and the statement view of science, we shall in the next chapter briefly look at the positivist statement account of science.

NOTES: CHAPTER 2

[1] Some sections of this chapter appear in Ruttkamp (1999a).

[2] Although not as common-sensical as Fine's (1986b) "natural ontological attitude" perhaps.

[3] One may ask how — and even if — it is possible to distinguish between conceptual and linguistic levels without giving a clear and valid answer to the question of whether it is possible to think without language. I am however not making rigid distinctions here. What I am doing, in fact, is to depict the development nf scientific research by emphasising one by one the real, conceptual, and linguistic aspects of this evolutionary process. And, moreover, I am claiming that there always is interplay between these aspects.

[4] Think, for example, of students able to cite all the rules (or laws) of a specific area of their subject matter, who are still unable to *apply* this knowledge in any concrete way.

[5] See for instance Balzer's articles "A logical reconstruction of pure exchange economics" (1982), and "The proper reconstruction of exchange economics" (1985); Hands's article "The structuralist view of economic theories: A review essay" (1985); and Janssen's article entitled "Structuralist reconstructions of classical and Keynesian macroeconomics" (1989); as well as Balzer and Hamminga's *Philosophy of economics* (1989), and many others.

[6] In a language such as L, we usually have the following eight categories of basic symbols available:
- a countable infinite set $\{v_i\}$ of individual variables
- a (possibly empty) set of individual constants
- a nonempty set $\{P_\alpha\}$ of predicate letters, and, associated with each predicate letter P_α, there is a positive integer $\delta(\alpha)$ called the arity of P_α, which gives the number of individual variables which are predicated by P_α
- a (possibly empty) set of function symbols
- the equality symbol "="
- logical connectives (details not important for my purposes here)
- quantifier symbols (ditto)
- punctuation symbols (ditto).

[7] A mathematical structure $U = <A, \{R_\alpha\}>$ consists of a set A, which is the domain of U, and a set of relations R_α (one for each α from some index set) defined on domain A. The sets A and $\{R_\alpha\}$ both may be infinite. A relation R_α on domain A is defined as a set of ordered $\mu(\alpha)$-tuples of elements from domain A, where $\mu(\alpha)$ is a unique non-negative integer associated with the relation R_α.

[8] In other words the mathematical structure U will count as an interpretation of the language L if and only if the arity of the relations R_α correspond to the arity of the predicate letters P_α. (That is, if $\delta(\alpha) = \mu(\alpha)$.) In this case U is called a realisation of the language L (and we can say that L is appropriate for the structure U). We call the relation R_α the value of P_α in the realisation U of language L. (If L has constant and function symbols, they are interpreted as elements of A and functions — of the proper arities — on A.)

[9] E.g., consider the formula Pxy. If P is interpreted as the relation < and if x and y are given the values of 3 and 5 respectively, then we say that Pxy is *true under that interpretation* and we say that formula Pxy in language L is *satisfied* by the valuation in the domain of interpretation U, ascribing the given values to variables x and y. (Because 3 is indeed smaller than 5.)

[10] Note that a realisation of language L is in principle a realisation of all the sentences in L, and this implies that every sentence in L is either true or false in that particular realisation.

[11] Einstein referred to these convictions as "free conventions" (Holton, 1995, p.464). "These themata, to which [Einstein] was obstinately devoted, explain why he would continue his work in a given direction even when tests against experience were difficult or unavailable (as in General Theory of Relativity), or, conversely, why he refused to accept theories well supported by phenomena, but, as in the case of Bohr's quantum mechanics, based on presuppositions opposite to his own, ..." (ibid., p.457).

[12] Chalmers (1993, p.202) gives another example of these events: "We may abstract the falling of [a] ... leaf from other aspects of its motion ... We then apply the appropriate fundamental laws [axioms of mostly "background" theories] to [this model] that [is] the result of our abstraction. We apply Newton's laws to the leaf as a mass subject to the gravitational attraction of the earth only, and derive the law of fall from it. Of course, since we have abstracted from winds, air resistance and the like, our model will not in general serve to describe the fall of any particular leaf. After all, the model is an abstraction. Nevertheless, provided we understand the leaf to have a capacity to fall, governed by Newton's laws, the *theoretical treatment via the abstract model* does explain the falling of the leaf, as distinct from its fluttering in the breeze". This is a point about which Nancy Cartwright has realist reservations, but which I still interpret as "realist" within a model-theoretic context. See Chapters 4 and 5 for more on Cartwright.

[13] Johannes Heidema (University of South Africa) introduced me to thinking in terms of this hierarchy.

[14] Einstein referred to the movement from conceptual structures or models to theories as a "creative leap" and in this sense referred to theories as "free creations of the human mind".

[15] Kepler's laws:
- First law: All planets follow elliptical orbits (and not circular ones, as Copernicus believed) with the sun situated in one of the foci of the ellipse.
- Second law: The line connecting the sun and a planet sweeps over equal areas of the planetary orbit in equal intervals of time.
- Third law: The squares of the periods of revolution of different planets around the sun stand in the same ratio (i.e. is proportional to) the cubes of their mean distances from the sun.

Newton's three laws of motion:
- Every body continues in its state of rest, or of uniform motion in a straight line, unless it is compelled to change that state by forces impressed on it.
- Change of motion is proportional to the force impressed, and is made in the direction of the straight line in which the force is impressed.
- The forces two bodies exert on each other are always equal and opposite in direction.

His law of gravitation:
- All material bodies attract each other with a force directly proportional to their masses and inversely proportional to the square of the distance between them.

[16] See the examples of the discoveries of Neptune and Pluto, as well as other applications of Newton's theory in the following section.

[17] Another type of approach to the interpretation of language terms is offered, for example, by Hans Lenk's methodological or schema interpretationism. See for instance Lenk (1993), (1995).

[18] Whenever I speak of models of theories, I am referring to the notion of model in the Tarskian sense that a model of a theory is an interpretation of the theory under which the set of sentences comprising the theory is true. As mentioned in the previous section, at the start of theory formulation the intended "model" scientists work with is not (initially) such a mathematical model, although at the stage of theory interpretation it becomes obvious that such (intended) models can be easily adapted such that they also are elements of the set of all (mathematical) models of the theory in question.

[19] These terms are the terms traditionally referred to as "theoretical" terms. Note that therefore I do not follow in the footsteps of advocates of the traditional version of the statement approach, in the sense that I do not need the kind of (too simple) distinction they make between theoretical and observational terms in the language of the theory. Rather than this forced division, I propose an approach in which theoretical and observational terms, as well as the difficult "correspondence rules" or "bridge principles" supposedly acting between these kinds of terms, all have natural interrelated non-unique and co-dependent roles to play at various levels of the scientific process. See Chapter 3.

[20] I claim that Giere's theoretical models, Wójcicki's theoretical and semantic models, and Suppes's physical and set-theoretic models are all mathematical models in this sense. Some of these authors make a similar kind of distinction that I make between these models as "intended" models — Wójcicki's theoretical models and Suppes's physical models — and these models as "interpretative models" interpreting the theory — Giere's theoretical models, Wójcicki's semantic models, and Suppes's set-theoretic models. See Chapter 4.

[21] This notion of changing the original set of assumptions made by the original scientist, naturally may be connected to Popper's theory of falsification.

[22] Note that in this case, the embedding function simply is the identitiy function, mapping elements of E_{emp} onto elements of E_{red}. See *Figure 2* below.

[23] Note that this distinction between so-called "theoretical" and "empirical" predicates is model-specific rather than unique or absolute. See Chapter 3.

[24] "... it is always legitimate for scientists to ask and sometimes possible for them to answer, questions about whether gasses are really composed of molecules or whether the earth really moves. Such questions cannot be rephrased as questions about the plausibility of our conceptions" (Bhaskar, 1978, pp.155). Well, model-theoretically, the verification of our interpretative models depends on being able to show how experiments concerning the data in question may be linked to these models (via certain empirical models). However, what Bhaskar means, I think, is rather that science does not determine the structure of reality, but rather discovers it.

[25] Formally, in Lipsey's (1983) terms, we have the demand function as $q_n^d = D(p_n, p_1, ..., p_{n-1}, Y, \xi)$, and the supply function as $q_n^s = S(p_n, F_1, ..., F_{n-1})$.

[26] Hausman's (1991, Chapter 3) term, in the sense of Giere's (1983), (1991) definition of "theoretical models".

[27] Now we can write the functions in endnote 25 as $q^d = a - bp$, $a,b \succ 0$, and $q^s = dp - c$, $c,d \succ 0$.

[28] Other examples: Models of neo-classical growth theory for instance are models of monetary growth, the one-sector model with exogenous population growth and technical progress, two-sector models, models with endogenous technical progress, and so on.

[29] See Torr (1999).

[30] Keuzenkamp and Magnus (1995) actually went so far as to challenge readers of the *Journal of econometrics* to "name a paper that contains significance tests which significantly changed the way economists think about some economic proposition" (ibid., p.21).

[31] Whether or not econometrics has an explanatory role — which is more important from a realist point of view — is a very difficult question, which can only be answered contextually.

[32] In the context of the Walrasian theory of tatonnement Patinkin (1965) asks who actually solves the equations. He (ibid., p.38) writes: "The fact that the number of independent excess-demand equations is equal to the number of unknown money prices and that the system can be formally solved, might some day interest a Central Planning Bureau duly equipped with ... computers and charged with setting equilibrium prices by decree. But what is the relevance of this fact for a free market functioning under conditions of perfect competition?".

[33] Think of caricatures exaggerating certain features of their subjects. If it is a good caricature, it is however possible to recognise the subject in question at a glance. (Chris Torr, Department of Economics, at the University of south Africa)

[34] Parts of this section forthcoming in the Poznan Studies's volume on Theo Kuipers (Amsterdam: Rodopi).

[35] More precisely, traditionally the nature of under-determination has been understood in terms of two kinds of relation between the "real world" and scientific theories. The first kind is taken to exist between phenomena (or whole systems) in reality and the observation terms of theories, while the second kind of relation is said to exist between sets of protocol sentences (formed from the observation terms and expressing data) and possible theories incorporating or explaining such a set of protocol sentences — that is, the existence of incompatible but empirically equivalent theories.

[36] Heidema and Burger (Forthcoming, p.1) note Paul's (1993) remark that abduction is often related to conjecture; diagnosis, induction, inference to the best explanation, hypothesis formulation, disambiguation, and pattern recognition.

[37] For instance: Clark's (1978) predicate completion, Reiter's (1980) default logic, McDermott and McDoyle's (1980) non-monotonic logic, McCarthy's (1980) circumscription, or McDermott's (1982) non-monotonic logic II. See also Ginsberg (1987), Kraus, Lehmann & Magidor (1990), and Shoham (1987).

[38] Where 'interpretation' means truth assignment for [propositional calculus], a first-order interpretation for [first-order predicate calculus], and a <Kripke interpretation, world>-pair for modal logic. (Ibid.)

[39] This example is borrowed from discussions with Willem Labuschagne from the Department of Computer Science at Otago University, Dunedin, New Zealand.

[40] There are two approaches to ordering possible worlds: by using numbers, or without using numbers. The best known numerical ways are those using *fuzzy sets* or using *probabilities*. Neither of these would give us the kind of formal mechanism I am looking for in the current context, and therefore I choose the mechanism of non-numerical default rules, applied by defining certain total pre-order relations to induce such orderings.

[41] It is natural to wonder whether a default rule can be expressed by a sentence of the formal language in the same way as our knowledge that, say, the light is on. The answer is no. A default rule, as we have construed it, says something about the ordering of states according to decreasing normality or typicality. Such default rules cannot be expressed as sentences of the logic language whose valuations or interpretations are intended to correspond to states of the system. That is why we cannot simply add in a default rule as an axiom and resort to familiar 'classical' logic. To see that default rules cannot be expressed as axioms, take a transparent propositional language for talking about the light-fan system. The predicate symbols are about the

components, expressing such notions as 'is on' or 'is malfunctioning'. The predicate symbols are not about the system as a whole. And if we add in constants denoting states and predicate symbols denoting relative normality, then interpretations of the new language would correspond to a new, more complex, system instead of to the old system.

[42] See Einstein, (1956, pp.11ff., 65ff.).

[43] Where i equals $\sqrt{-1}$.

CHAPTER THREE

THE STATEMENT ACCOUNT OF SCIENCE

1. INTRODUCTION

At least the fact that the model-theoretic notion of a scientific theory as a deductively closed set of sentences is also a feature of theories accepted by the advocates of the statement account of science seems to imply that model-theoretic realism might overlap with their views also in other ways. Well, for one thing, a defender of model-theoretic account of science also looks to a semantic analysis of scientific language to fulfil her goals. Defenders of model-theoretic *realism*, however, also want to be realists, and so they want something better (or "more") than Carnap's (1966, p.256) rather non-committal view of the instrumentalist-realist debate:

> My own view ... is that the conflict between the two approaches is essentially linguistic. It is a question of which way of speaking is to be preferred under a given set of circumstances. To say a theory is a reliable instrument — that is, that the predictions of observable events that it yields will be confirmed — is essentially the same as saying that the theory is true and that the theoretical, unobservable entities it speaks about exist.

Nagel also seems to be sympathetic to such a portrayal of the realist-instrumentalist debate. He (1961, p.139) writes:

> Moreover, as has already been suggested, questions can be raised about a theory when it is regarded as a leading principle [instrumentalism] that are substantially the same as those which arise when the theory is used as a premise [realism]. For whether or not a material leading principle happens to be a theory, the principle is a dependable one only if the conclusions inferred from true premises in accordance with the principle are in agreement with facts of observation to some stipulated degree. In consequence, there is on the whole only a verbal difference between asking whether a theory is satisfactory (as a technique of inference) and asking whether a theory is true (as a premise).

(Empirical) scientific theories may indeed be reconstructed as positivist logical structures of (created) statements, accountable to empirical input through their predictive consequences which are identifiable by observation (if this latter notion is given its broadest interpretation). Checking this accountability is however not a mere empirical matter but also a semantic one. Tying logical analyses to exact — in the sense of "unique" — determinations of the meaning of linguistic expressions cannot succeed, given the fact that these determinations are contingent on the nature of the very models they help define. I shall elaborate on this in what follows, but first, let us consider more closely the so-called statement account of science offered by the logical positivists.

2. THE TWO-LANGUAGE VIEW OF SCIENTIFIC THEORIES

I shall first briefly discuss the "two-language" aspect of the positivist statement account of science best depicted by the logical positivists of the Vienna Circle (such as Carnap, Nagel, Feigl, Braithwaite, Reichenbach, Campbell, Ramsey, and many others). The supporters of the positivist-statement account of science depict the rational reconstruction of the language of science as a syntactic system with an axiomatised deductive theory formulated within that system. Hempel (in Suppes et al. (1973, p.368)) explains that usually defenders of this view divide a theory analytically into two constituent classes of sentences. First we have the set of so-called "internal principles" of the theory which describes "underlying entities and processes postulated by the theory" and states the "laws" or "theoretical principles" that are taken to govern them. This set of internal principles is a set of formulas containing uninterpreted extra-logical constants (that is, the so-called theoretical terms), which if axiomatised, form an uninterpreted axiomatised formal system or calculus. The extra-logical constants are composed of both undefined terms and certain defined terms. The so-called theoretical terms (which in this case are terms that do not have obvious relations with "observation" terms, i.e. terms like "electron", "particle", "mass", and so on) are thus set out in an abstract formal calculus (i.e., in a symbolic language). Nagel (1961, p.90) characterises this abstract calculus as the logical structure of the "explanatory system" (i.e. the theory), and claims it to "implicitly define the basic notions of the system".

The positivist depiction of science traditionally required that every theoretical term in a scientific theory be provided with an "explicit definition" composed entirely of observational terms. The second set of sentences identified in the statement view thus is such a set of explicit definitions. It might be noted here that, although Carnap, for instance, had already by as early as 1936 given up on the notion of explicit definitions, this notion is of sufficient importance to the key notions of theory structure and reference to deserve relatively lengthy discussions in what follows. Hempel refers to them as "bridge principles" (Hempel in Feigl et al. (1958, p.46)), but I shall refer to them as correspondence rules, or c-rules, following Carnap. This set of sentences "... indicate[s] ways in which occurrences at the [theoretical] level ... are held to be linked to the phenomena the theory is to explain" (ibid.). The sentences in this set of c-rules are "... viewed as affording interpretations of theoretical expressions in a vocabulary whose terms have fully determinable and clearly understood empirical meanings ..." (ibid.). This set of correspondence rules assigns empirical (observational) content to the logical calculus by providing "co-ordinating definitions" or "empirical interpretations" for at least some of the theoretical terms in the logical calculus in terms of "pure" observational terms, i.e. the c-rules are said to relate the abstract calculus to "the concrete materials of observation and experiment" (Nagel, 1961, p.90). Hempel (in Feigl et al., 1958, p.47) remarks that

> ... indeed, for the deductive development of the axiomatised system, no meanings need be assigned at all to its expressions, primitive or derived. However, a deductive system can function as a theory in empirical science only if it has been given an *interpretation* by reference to empirical phenomena. We may think of such interpretation as being effected by the specification of a set of *interpretative sentences*, which connect certain terms of the theoretical vocabulary with observational terms.

Nagel (ibid.) also identifies a third component of theories, which he describes as "... an interpretation or a model for the abstract calculus, which supplies some flesh for the skeleton structure in terms of more or less familiar or visualisable materials". What Nagel has in mind here is much closer to a kind of Hesse-ian analog model than to the mathematical Tarskian structures I call models (see also Chapter 1). Notice that all three of the above components of theories are formulated within a "... linguistic framework of a clearly specified logical structure, which determines, in part, the rules of deductive inference" (Hempel in Feigl et al. (ibid., p.46)).

Now, in the context of our discussion in terms of model-theoretic semantics, the positivist claim is thus (Przelecki, 1991, p.287) that observational terms are assigned their interpretation(s) independent of the theory. Theoretical terms, in their turn, however, "acquire [their interpretation(s)] through their connections with the observational terms as established by theory T (or rather by its definitional, analytic component)" (ibid.) — i.e. the set of correspondence rules. "The observational terms are thus said to be interpreted directly by ostensive (or operational) procedures, [and] the theoretical terms indirectly, by postulates [definitions] relating them to the former" (ibid.). Both kinds of interpretational procedure are problematic. As Przelecki (ibid.) claims, the interpretation of observational terms is indeterminate because of their vague ostensive nature, but also, more importantly, because the positivist procedure assigns in fact a whole class of interpretations to an observational term, each of which corresponds to a possible way of making the term precise. Theoretical terms "inherit" (ibid.) this indeterminacy because their interpretational procedure implies that they are interpreted through postulates (the c-rules) connecting them with the observational terms. Furthermore these (c-)postulates are themselves indeterminate. (It is in this sense that one may refer to the "openness" (ibid.) of theoretical terms.) Before we elaborate on the problematic nature of c-rules, the following.

Carnap (1966, p.258) divides the terms in a scientific language into three main groups: logical terms (including all the terms of pure mathematics), observational terms, and theoretical terms (which are "constructs" (ibid.)). He offers an analogous division of the sentences of a scientific language: Logical sentences which contain no descriptive terms; observational sentences which contain observational terms but no theoretical terms; and theoretical sentences which may be of two kinds: mixed sentences containing both theoretical and observational terms, and purely theoretical sentences containing only theoretical terms. In this way he claims that the "... entire language of science is conveniently divided into two parts. Each contains the whole of logic (including mathematics). They differ only with regard to their descriptive, nonlogical elements". He (in Hintikka (1975, p.75)) claims that the

> ... observational language contains only an elementary logic. The sentences of this language are assumed to be fully understood. The meaning of the theoretical language, on the other hand, always remains incomplete. It possesses a very comprehensive logic which contains all of classical mathematics.

So, in this sense, the observational language (L_O) contains observational sentences, but no theoretical terms, while the theoretical language (L_T) contains logical and theoretical sentences, with or without observational terms. In this way, it is understood that theoretical terms

> ... are introduced into the language of science by a theory, T, which rests upon two kinds of postulates — the theoretical, or T-postulates, and the correspondence, or C-postulates. The T-postulates are the laws of the theory. They are pure theoretical sentences. The C-postulates, the correspondence rules, are mixed sentences, combining theoretical terms with observational terms. (Carnap, 1966, pp.258, 259)

In other words, (Carnap in Hintikka (1975, p.76)) the languages L_O and L_T are sublanguages of the scientific language L such that the constants of L are divided into logical and nonlogical (or descriptive) constants. (All mathematical constants are logical.) The primary descriptive constants of L_T are theoretical terms (which are terms not explicitly definable by observational terms), while the primary descriptive terms of L_O are observational terms.

The sublanguage L_T is obtained by adding descriptive theoretical terms to the mathematical language via theoretical postulates (the "laws" of the theory in question) containing only T-terms and correspondence postulates containing both T- and O-terms. Carnap (in Hintikka (ibid., p.79)) remarks that

> For the assertion of facts and the formulation of laws, in any empirical scientific theory, descriptive terms are obviously necessary. In this, it seems to me the old saying that the book of nature is written in the language of mathematics is quite misleading.

Simultaneously though, he (ibid., pp.80, 81) points out that

> ... it is not necessary to assume new sorts of objects for the descriptive T-terms of theoretical physics. These terms designate mathematical objects ... which, however, are physically characterised, ... so they have the relations to the observational processes established by the C-postulates while simultaneously satisfying the conditions given in the T-postulates.

This almost reminds one of the distinction between linguistic, conceptual (interpretative), and empirical levels of science as portrayed in Chapter 2.

3. THEORIES AND NON-OBSERVABLES

Let us now focus a little closer on how exactly two of the main figures of the positivist, statement, or "received" view (namely Rudolf Carnap and Ernest Nagel) define the above concepts. Contributing to the theoretical/observational distinction, is the traditional statement view distinction between empirical laws and theoretical laws. Carnap (1966, p.225) defined empirical laws as "... laws that can be confirmed directly by empirical observations", and he explains that since the term "observable" "... is often used for any phenomenon that can be directly observed, ... it can be said that empirical laws are laws about 'observables'" (ibid.).

To Carnap the distinction between what can be termed as an observable, and what as a non-observable, has to be made according to whether an object is directly observable by the senses or measurable by relatively simple techniques (this would be an observable object), and objects that need complex and indirect procedures for measurement (unobservables) (ibid., pp.225 - 226). Obviously, these points of distinction can not be regarded as offering us any final criterion for observability.

Carnap (ibid., p.226) remarks that "There is a continuum which starts with direct sensory observations and proceeds to enormously complex, indirect methods of observation. Obviously no sharp line can be drawn across this continuum; it is a matter of degree".

Nagel (1961, p.80) defines what he calls an "experimental law" as formulating

> ... a relation between things (or traits of things) that are observable ["experimentally observable" (ibid., p.81)] in the admittedly loose sense of 'observable' ... and ... the law can be validated (even if only with some 'degree of probability') by controlled observation of the things mentioned in the law.

In his context, Carnap (1966, pp.226, 227) adds that these (empirical) laws generalise the results of observation and measurement and are therefore sometimes referred to as empirical generalisations. — "The scientist makes repeated measurements, finds certain regularities, and expresses them as a law. These are empirical laws ... they are used for explaining observed facts and for predicting future observable events" (ibid.).

Theoretical laws, on the other hand, are laws that contain terms of a different nature than the terms contained in empirical laws. Carnap (ibid., p.228) states that

> The terms of a theoretical law do not refer to observables even when the physicist's wide meaning for what can be observed is adopted. They are laws about such entities as molecules, atoms, electrons, protons, electromagnetic fields, and others that cannot be measured in simple, direct ways.

Nagel (1961, p.82), in his turn, writes that

> ... though the commonly cited examples of theories are statements about things that in an obvious sense are unobservable, it is frequently possible to determine indirectly, by way of inferences drawn from experimental data in accordance with certain rules, important characteristics of what is ostensibly not observable.

This implies Carnap's (1966, p.228) point that theoretical laws are not generalisations of empirical laws, or an abstract way of capturing regularities among empirical generalisations. Rather, as Carnap (ibid., p.229) writes, theoretical laws help explain empirical laws and predict "new" not as yet observed facts.

> Just as the single, separate facts fall into place in an orderly pattern when they are generalised in an empirical law, the single and separate empirical laws fit into the orderly pattern of a theoretical law. This raises one of the main problems in the methodology of science. How can the kind of knowledge that will justify the assertion of a theoretical law be obtained? An empirical law may be justified by making observations of single facts. But to justify a theoretical law, comparable observations cannot be made because the entities referred to in theoretical laws are nonobservables.

Thus, notice that theoretical laws, rather than stated as generalisations of "particular concrete facts that can be spatiotemporally specified" (ibid.), are stated as hypotheses. How do we "test" theoretical hypotheses though? We derive some empirical laws from the hypothesis and test these empirical laws by observation of facts. If such "derived" empirical laws are observationally "confirmed", indirect (or partial) confirmation of the theoretical law in question has been provided. Nagel (1961, p.79) affirms this position when he writes that

> Scientific knowledge takes its ultimate point of departure from problems suggested by

68   CHAPTER 3

observing things and events encountered in common experience; it aims to understand these observable things by discovering some systematic order in them; and its final test for the laws that serve as instruments of explanation and prediction is their accordance with such observations.

Now Nagel (ibid., pp.83ff.) offers us some "principles" for distinguishing between empirical and theoretical laws. First, he claims that each "'descriptive' (i.e. nonlogical) term" in an experimental law is associated with "... at least one overt procedure for predicating the term of some observationally identifiable trait when certain specified circumstances are realised" (ibid., p.83), while in general this is not the case for each such term of a theoretical law. From this Nagel deduces that thus "... an experimental law, unlike a theoretical statement, invariably possesses a determinate empirical content which in principle can always be controlled by observational evidence obtained by those procedures", and thus he concludes (ibid.) that the procedure associated with terms in experimental laws ascribes definite, although partial, meanings for these terms such that empirical laws can be tested in the light of data acquired by way of these procedures.

The second point of distinction Nagel offers is that experimental laws could in principle be "proposed and asserted as individual generalisations based on relations found to hold in observed data" (ibid., p.85), which is not the case for theoretical laws. This should be hardly surprising, and is also Carnap's point above: a theory cannot be an empirical generalisation from empirical data for the obvious reason that "... in general there are no experimentally identifiable instances that fall into the scope of predication of the theory" (Carnap, 1966, p.229). In this context Nagel (1961, pp.85, 86) makes the following very important point about the nature of theories:

> Distinguished scientists have repeatedly claimed that theories are 'free creations of the human mind'. Such claims obviously do not mean that theories may not be *suggested* by observational materials or that theories do not require support from observational evidence. What such claims rightly assert is that the basic terms of a theory need not possess meanings which are fixed by definite experimental procedures, and that a theory may be adequate and fruitful despite the fact that the evidence for it is necessarily indirect.

As a defender of model-theoretic realism I agree, and add that besides being indirect, this "evidence" may also be amendable (see Chapter 2 again).

Nagel's point is that the history of science shows that theories are often accepted solely on the basis that they can explain already established experimental laws. This leads us to a third (related) principle of distinction between theoretical and experimental laws. Even in cases where an experimental law is explained by a certain theory and forms a part of the theory's framework, two features of the experimental law remain unchanged: it "retains a meaning that can be formulated independent of the theory" and it "is based on observational evidence that may enable the law to survive the eventual demise of the theory" (ibid., p.86). Theoretical terms, on the other hand, can only be understood as part of the particular theory in which they appear. The reason for this is that

> ... theoretical terms are not assigned a unique set of determinate senses by the postulates of a theory, the permissible senses are limited to those satisfying the structure of interrelations into which the postulates place the terms. Accordingly, even when the final

postulates of a theory are altered, the meanings of its basic terms are also changed, even if (as often happens) the same linguistic expressions continue to be employed in the modified theory as in the original one. The new theory will presumably continue to explain all the experimental laws that the earlier theory could explain, in addition to explaining experimental laws for which the earlier theory could not account. But in consequence of the changed theoretical content of the new theory, the observationally identifiable regularities that are formulated by experimental laws and explained by both the original and the modified theory receive what are in fact different theoretical interpretations. (Ibid., p.87)

All of which is accommodated model-theoretically within a realist context. (See Chapters 2 and 5.)

A last point of distinction (ibid., p.88) is the fact that theories are usually presented as systems of related statements, while experimental laws are formulated in single statements. Apart from the fact that this is in accordance with the fact that theories provide "fresh" (ibid., p.90) suggestions for experimental laws, it is also in accordance with the fact that "... theoretical notions are not tied down to definite observational materials by way of a fixed set of experimental procedures" and also "... because of the complex symbolic structure of theories more degrees of freedom are available in extending a theory to many diverse areas" (ibid., p.89).

4. RULES OF CORRESPONDENCE

So, if it is the case, as implied by the above, that logical empiricists portray "theoretical knowledge in empirical science as grounded on the data of direct observation in ways made explicit by correspondence rules" (Hempel in Suppes et al. (1973, p.371)), let us look closer at the nature of this set of rules. Nagel (1961, p.93) writes that "... if a theory is to explain experimental laws, it is not sufficient that its terms be only implicitly defined", since we have to "indicate how its implicitly defined terms are related to ideas occurring in experimental laws" (ibid.). But the theoretical postulates of the theory do not (cannot, since they are not defined that way) offer us any information about the nature of the subject matter or observable material to which the theory can be applied, since these postulates do not specify the substantive character of any *specific* system, but only the logical skeleton of *some* system.

However the idea of c-rules as explicit definitions soon showed itself to be untenable. Keep in mind that an expression explicitly defined may at any time be eliminated from any context in which it occurs because it can be replaced by the defining expression without altering the sense of the context, which implies that the defining expression and the expression defined are equivalent in meaning. Carnap (1966, pp.234, 235) writes:

> There is a temptation at times to think that the set of [correspondence] rules provides a means for defining theoretical terms, whereas just the opposite is really true. A theoretical term can never be explicitly defined on the basis of observable terms, although sometimes an observable can be defined in theoretical terms. ... There is no answer to the question: 'Exactly what is an electron?' ...[But] [t]here is no special mystery here. There is only an improperly phrased question. Definitions that cannot, in the nature of the case, be given, should not be demanded. ... we do not possess a picture of an electron. ... a physicist can describe the behaviour of an electron only by stating theoretical laws, and these laws contain theoretical terms. ... There is no way that a theoretical concept can be defined in terms of observables. We must, therefore, resign ourselves to the fact that definitions of the kind that can be supplied for observable terms cannot be formulated for theoretical terms.

This is acceptable, and more than that, actually the case, as we now know. This does however, not imply anything negative regarding realism. All that it implies for defenders of realism is that the nature of c-rules, or of the process of "linking" theoretical terms with observational ones, must be carefully redefined in the interest of tracing relations of reference between theories and empirical reality.

Hempel (in Feigl et al. (1958, pp.50ff.) also discusses the possibility of characterising c-rules as explicit definitions. He (ibid., p.50) writes:

> In the simplest case, such a sentence could be an *explicit* definition of a theoretical expression in terms of observational ones ... . In this case, the theoretical term is unnecessary in the strong sense that it can always be avoided in favour of an observational expression, its definiens. If all the primitives of a theory T are thus defined, then clearly T can be stated entirely in observational terms, and all its general principles will indeed be laws that directly connect [non-] observables with observables.

Hempel (ibid.) thus rather proposes an operational definition for c-rules: "Def.$Q(x) \equiv (Cx \supset Ex)$", which implies that (by definition) a certain object x has the property Q if and only if it is such that if it is under test conditions of kind C, then it exhibits an effect of kind E. Carnap, however has shown (Carnap, 1936; 1937) that these kinds of definitions of c-rules have a serious flaw: Recall that a conditional sentence is false only if its antecedent is true and its consequent is false. Considering a case where the antecedent of Hempel's conditional sentence is false, that is, if we consider an object which does not satisfy the test conditions denoted by C, we still find the definiens as a whole to be true, and that the object in question may be assigned the property Q (regardless of the fact that it does not satisfy conditions C).

Carnap proposes that we have to provide a partial specification of the meaning for "Q", rather than a complete one. He does this by means of so-called "reduction sentences". He replaces the above with the reduction sentence $Cx \supset (Qx \equiv Ex)$. This sentence implies that if an object x is under test conditions C, then it has the property Q if and only if it exhibits an effect of kind E. Here, if our object is not under test conditions C, then the entire formula that Carnap offers is still true of it, but this implies nothing as to whether the object does, or does not, have the property Q. Carnap's proposal specifies only a partial meaning for Q (and not a complete explicit (albeit operational) definition as in Hempel's case), since Q is only meaningful in cases where the object under consideration meets conditions C.

Hempel (ibid., p.52) comments that

> By construing the latter as merely partial specifications of meaning, this approach treats

theoretical concepts as 'open'; and the provision for a set of different, and mutually supplementary, reduction sentences for a given term reflects the availability, for most theoretical terms, of different operational criteria of application, pertaining to different contexts.

Notice thus that the notion of reduction sentences also confirms the fact that theoretical terms cannot be fully defined, or specified, or given their full meaning, by reference to observables. Both of the above points are, at least in spirit, what is implied by a model-theoretic characterisation of reference relations too (Chapter 2).

We do however, as pointed out earlier need some form of "connection", however temporary in my terms and however defeasible, between theories and reality. This is the case for realists but is also acknowledged by Carnap, for instance. He (Carnap, 1966, p.237) claims that this is one important difference between an axiomatic system in mathematics and one is physics, in the sense that a mathematical term can be interpreted by a definition in logic, and also such an interpretation is final, none of which is the case with a physical term. This is why axiomatic terms in physical systems (such as "electron" and "field") "... must be interpreted by correspondence rules that connect the terms with observational phenomena" (ibid., p.237). Also such interpretations are necessarily incomplete, and the "system is left open to make it possible to add new rules of correspondence" (ibid.) whenever for instance new procedures are developed for measuring observable terms. Thus interpretations for theoretical terms can always be amended in the sense of new c-rules "increasing the amount of interpretation specified" (ibid., p.238) for these terms. And so, should we ever wish to speak of a "complete" or "final, explicit" (ibid.) definition for some theoretical term, we would actually be speaking of a theoretical term that has become part of the observational language and which thus has ceased to be theoretical (as pointed out in Hempel's terms above).

This reminds me of the tension between realism and empiricism that I shall comment on in Section 5.7. We need some empirical input for validating the articulation of the relations (however they are defined) between theories and reality. This is also the sense in which Carnap wants us to see philosophy (of science) demarcated from metaphysics. He (ibid., p.245) writes:

> We are on solid ground ... if we issue the warning that no hypothesis can claim to be scientific unless there is the *possibility* that it can be tested. It does not have to be confirmed to be a hypothesis, but there must be correspondence rules that will permit, in principle, a means of confirming or disconfirming the theory.

In Hintikka (1975, p.78) he comments "Every physicist is aware that perhaps tomorrow a lack of correspondence will be discovered. He [sic] is prepared ahead of time in this case, to modify his [sic] system. ... There is no ground for despair". Indeed, there is not, as I show by my arguments about non-monotonicity (in terms of preference relations) concerning "temporary empirical" knowledge in the sense of fast changing data regarding the empirical world and the proliferation of empirical models. So, clearly, a second reason for not viewing c-rules as explicit definitions, is the fact that theoretical notions are only implicitly defined by the relevant theory's postulates (even if the theory is "presented by way of a model" (Nagel, 1961, p.99)), which may imply what I refer to as "over-determination" of theories by empirical models (Chapter 2). In line with this Nagel (ibid., p.106) writes that

> ... theoretical notions are frequently coordinated by rules of correspondence with more than one experimental concept. ... There are therefore an unlimited number of experimental concepts to which, as a matter of logical possibility, a theoretical notion may be made to correspond. ... Accordingly, in those cases in which a given theoretical notion is made to correspond to two or more experimental ideas (though presumably on different occasions and in the context of different problems [sometimes in new unexpected contexts]), it would be absurd to maintain that the theoretical concept is explicitly defined by each of the two experimental ones in turn.

Yes, and listen to this:

> The haziness that surrounds ... correspondence rules is inevitable, since experimental ideas do not have the sharp contours that theoretical notions possess. This is the primary reason why it is not possible to formalise with much precision the rules (or habits) for establishing a correspondence between theoretical and experimental ideas. (Ibid., p.99)

These aspects of coordinations between theories and real systems are also at home in the model-theoretic context in the sense that one theory may have different empirical reducts (different theoretical terms may be given "empirical interpretations" at different times, mainly because of changes in the body of scientific and technological knowledge available in a specific disciplinary matrix at a given time), together with the fact of more than one empirical model being isomorphically embedded into such reducts. But model-theoretically, we can still be realists (not "traditional" ones, no, but that's okay, we do not want to be limited in that way anyway, and also regardless of whether Carnap and Nagel did or did not care to be).

Rounding up this section, we must briefly mention Hempel's notion of "antecedently available" knowledge. In Suppes et al. (1973, pp.372, 373) Hempel writes that "... the requirement of an observational interpretative base for scientific theories is unnecessarily artificial". He (in Feigl, 1958, p.83) claims that "... we might qualify a theoretical expression as intelligible or significant if it has been adequately explained in terms which we consider as antecedently understood". Thus Hempel construes the interpretative base of a theory as consisting of antecedently available predicates, rather than of observational predicates. Important also, is that because this notion is obviously a relative one, or of an "historic-pragmatic" (in Suppes, 1973, p.373) nature, it "... affords a plausible construal of the public and inter-subjective character of the evidential base for scientific theories by linking it to the uniformity with which the antecedent vocabulary is used by scientists trained in the field" (ibid.). Hempel's notion of antecedent specification is also echoed in the model-theoretic context where it is the specification of empirical reducts of models which decides positivist "theoretical/observational" distinctions (in the context of particular empirical interpretations), such that these distinctions are *model*-specific and also not absolute at all.

In addition to the problems with the statement account of c-rules discussed above, I offer the following two problems which coincides with two of the reasons Wójcicki (1996, pp.85ff.) offers why formal reconstructions of empirical theories encounter problems.

- There are empirical quantities which cannot be adequately characterised within a single theory (such as the notion of energy (ibid.)) and so certainly cannot be "coordinated" with "observational terms" in a unique way.

- There are empirical principles which cannot be adequately stated within one theory (such as the principle of invariancy of physical laws under Lorentzian transformations (ibid.)).

All of the above indeterminacies concerning theoretical/observation distinctions and the nature of c-rules led to a number of views on the nature of c-rules and the theoretical/observational distinction that oscillate between being almost naively optimistic or utterly cynical. The requirement for a set of c-rules (or "postulates") to connect theoretical terms to their observational counterparts was supposed by some to be the tool for actualising the positivist dream of rooting out all forms of pseudo science, but, in a sense, turned into the biggest enemy of the positivist ideal. Briefly, the reason for this is that it is impossible — given all of the above — to find one clear unambiguous method in which to draw the observational/theoretical distinction, mainly because of the spurious nature of the positivist definition of c-rules. Wójcicki (ibid., p.86) makes the same point: —

> Suppose that a deductive system is furnished with an empirical interpretation. Under the positivist conception of science, an empirical interpretation consists in reducing theoretical terms to observational ones. Different definitions of how this reduction is to be carried out were proposed at different periods of time. There are not many theorists of science who would still be ready to subscribe to the idea of the observational-theoretical dichotomy. But the reduction of theoretical terms to observational ones is but one of many ways in which deductive systems can be supplied with an empirical interpretation. The simplest and the most natural way to interpret empirically mathematical statements is to relate mathematical quantities they involve to appropriately selected measurement procedures combined with techniques of processing empirical data.

The field of proposals for the determination of c-rules stretches from Ramsey's formulation of special existentially quantified sentences, as well as Carnap's (initial) criterion in terms of the relative ease with which a scientist can determine a certain theoretical term"t"'s correct application and his later reworking of Ramsey sentences, to Grover Maxwell's (1962) famous "The ontological status of theoretical entities" in which he discusses the variety of interpretations of the notion "observable", to the so-called instrument criterion (whether an "artificial" instrument is needed to apply "t" or not), to Putnamian cynicism (no non-arbitrary distinction possible), to the negative implications for a unique t-distinction coming from Quine's and Duhem's holistic web of belief view, to David Lewis's (1970) suggestions in terms of defining theoretical terms by an application of Carnap's use of Ramsey sentences, and so on and so on. This implies that the defenders of positivism cannot claim that using formalised c-rules expressed in the language of first-order logic provides us — necessarily — with a unique and precise understanding of the relation between theoretical terms and observational terms.

Against this background, the question whether the tools of model theory may be of help to define the T-distinction more clearly and to get a grasp on the non-unique character of c-rules themselves, becomes rather interesting. Nagel (1961, p.91), for instance (see Chapters 1 and 2) — depending on his interpretation of the notion of "model" — may be denying the possibility of adequately using model theoretic semantics with these objectives in mind, since he denies that correspondence rules may be — or are — given by models or interpretations of theories. He acknowledges that the latter structures may give "meaningful statements" (ibid.) which may turn out to be explicit definitions for terms in theories, but points out that correspondence rules offer "rules for coordinating [the theory's] nonlogical expressions [i.e. its theoretical terms] with *experimentally significant notions*". This seems to imply then that models do not have the ability to give "observational" content to terms in theories. This might be the case — although model-theoretically it is not — but anyway does not necessarily imply that semantic analyses of science cannot offer us refinements of the nature of the T-distinction.

In the next chapter we shall take a closer look at some non-statement accounts of science, which shall enable us among other things, in the end, to determine whether — if at all — these kinds of analysis of theories offer perhaps a more fertile interpretation of correspondence rules than the positivist interpretation there-of. For now, just a few words on non-statement sentiments on the nature of c-rules. According to Suppes (1967, p.57) a problem with the statement approach's depiction of correspondence rules is that they do not in the sense of modern logic offer an adequate semantics for the axiomatic calculus of the theory. Wójcicki (in Humphreys, 1994, p.127) explains that

> ... if for the logical positivists the right way to define an empirical theory T was to define a set of axioms from which all the other sentences valid in T are logically derivable, Suppes suggests that to define T is to define a set-theoretical predicate that denotes all the set-theoretical structures [semantical models] of which T is true in the Tarski sense.

Suppes does not so much emphasise the non-statement approach versus the statement approach, although he implies that his set-theoretical approach does offer an adequate semantics for the axiomatic calculus of a scientific theory. In other words, he seems to indicate that his set-theoretical semantics excludes the need for c-rules in the "statement" context, since he discards the need for some appropriate language in which to formulate theories. He (Suppes, 1954, p.244) writes:

> ... Why axiomatise? ... a good many philosophers seem to labour under the misimpression that to axiomatise a scientific discipline ... one needs to formulate the discipline in some well-defined artificial language. ... this kind of linguistic viewpoint is, in my opinion, seriously in error, and the predominance of this attitude has perhaps been one of the major reasons for the lack of substantial positive results in the philosophy of science. ... . The viewpoint I am advocating is that the basic methods appropriate for axiomatic studies in the empirical sciences are not metamathematical (and thus syntactical and semantical) but set-theoretical[1].

A problem that has been worrying Cartwright (1983) too — and which occupies any philosopher concerned with dealing with the intricacies of realism, maybe especially those working from a model-theoretic point of view — is, very simply put, that the presence or necessity of c-rules, however they are interpreted (in terms of models, mathematical functions, both, or something entirely different), implies somehow that the theory[2] itself has very little to say about the real phenomena the c-rules are supposed to link it to. First on her mind in *How the laws of physics lie* (Cartwright, 1983, Chapter 8), is to make clear whether having as few as possible c-rules should hold a promise of high explanatory power. I suppose the reasoning behind this question is that the fewer c-rules or bridge principles a theory needs the less "fundamental" — in the sense of falsity — its fundamental laws. She (ibid., p.143) claims that, if members of some research community want to be able to work together, some way has to be found in which to limit the kinds of models that can be used to describe real phenomena, because, given the complex and rich nature of these objects (the phenomena that have to be described) a variety of models might describe the same phenomena. She (ibid., p.144) writes :

> If there were endlessly many possible ways for a particular research community to hook up phenomena with intellectual constructions, model building would be entirely chaotic, and there would be no consensus of shared problems on which to work. The limitation on bridge principles provides a consensus within which to formulate theoretical explanations and allows for relatively few free parameters in the construction of models ... It is precisely the existence of relatively few bridge principles that makes possible the construction, evaluation, and elimination of models. This fact appears also to have highly anti-realist side effects ... it strongly increases the likelihood that there will be literally incompatible models that all fit the facts so far as the bridge principles can discriminate ... .[3]

I do not think that limiting the number of c-rules or bridge principles will solve Cartwright's problem. I am comfortable with the more "natural" limitation on models provided by the aims, background information, equipment, and training of the specific scientific community in question, as well as the satisfaction functions operating between theories and their models that set out the specific boundaries of the content of models in the first place. Related, and perhaps most importantly, the amount of control (in terms of *articulating* reference) that we get by turning towards preferential or minimal model semantics (Chapter 2) when faced with many models of one theory is enough to satisfy my realist objectives.

Thus, as far as c-rules or bridge principles are concerned — they do *not* hold *ceteris paribus*. There is no absolute set of rules describing these kinds of correspondence relation. Rather these rules hold with respect to a certain model within whose boundaries the theory is true. The best that can be said about "bridge principles" in my terms is that perhaps one can speak of a set of bridging "procedures" or "links" that extracts data from the relevant real system relative to a specific empirical context, and then injects these data, as an empirical model into the model under consideration.

Perhaps the most important thing there is to say about the supposed theory/observation "split" is that this should be described at most as a((n) inclusive) disjunction. The language of science can be observational or it can be theoretical or it can be both, without in any of these three cases smashing any hopes of tracing reference to "something" in some system of reality. How can this be? Well, precisely since the observational-theoretical distinction is a decision that is contingent on the model within which the theory we are focussing on is interpreted. The conditions under which such a model makes the relevant theory true, will be the ones that determine the formulation of the empirical reduct we formulate or extract as it were from the interpretative model itself. And it is enough for the purposes of tracing reference to agree to call the terms in such a reduct (under that particular empirical interpretation) observational or "empirical" and to refer to the ones that we do not include but that are still interpreted in the interpretative model as "theoretical".

In other words, the definition of c-rules remains somewhat less precise than one might wish for. This point is illustrated by Nancy Cartwright (1994b, p.292) when she claims "... we build our circumstances to fit our models ... it is just the point of scientific activity to build models that get in, under cover of the laws in question, all and only those circumstances that the laws govern". In other words, scientific theories may indeed say something about reality (because of the way the reality-model-theory link is interpreted), but it is not possible when faced with an *uninterpreted* theory to determine or claim that it will definitely be applicable to a certain aspect of reality and to no other.

The history of philosophy of science has shown though, that non-statement accounts of science have devised more and more precise ways to trace and define specific theoretical/observational distinctions. (See also Chapter 4.) From the structuralist constraint on the class of potential models of a theory to the class of partial potential models, to the model-theoretic notion of articulated reference and truth, the nature of c-rules has been made expressible to the finest detail with the aid of the semantic notion of models (see also my comments on the implications for correspondence links of my non-monotonic (in terms of a minimal model semantics) treatment of empirical proliferation in Chapter 2). Even Cartwright (1989) claims scientific theories to refer — albeit to what she terms causal capacities — if and only if the set of models in which the theory in question is true, is such that it has some further class of substructures where imputations of causal capacities are concerned, and is coextensive with the set of models actually used as working interpretations of the theory in question. More about these views in the next chapter.

I give Hempel the last word on c-rules and the theory/observation distinction, before we briefly turn to Ramsey sentences and related problems, after which, finally, we shall move on to variations of the non-statement view in Chapter 4. Hempel (in Feigl et al., 1958, pp.49,50) describes what he refers to as the (by now well-known) "theoretician's dilemma". The dilemma is defined (ibid.) as follows:

> ... if the terms and general principles of a scientific theory serve their purpose, i.e., if they establish definitive connections among observable phenomena, then they can be dispensed with since any chain of laws and interpretative statements establishing such a connection should then be replacable by a law which directly links observational antecedents to observational consequences. [This implies that] ... if the terms and principles of a theory serve their purpose they are unnecessary, ... and if they don't serve their purpose they are surely unnecessary. But given any theory, its terms and principles either serve their purpose or they don't. Hence, the terms and principles of any theory are unnecessary.

Instead of solving the dilemma, he (ibid., p.87) concludes though that

> ... when scientists or methodologists claim that the theoretical terms of a given theory refer to entities which have an existence of their own, which are essential constituents or aspects of the world we live in, then, no matter what individual connotations they may connect with this assertion, the reasons they could adduce in support of it seem clearly to lie in the fact that those terms function in a well-confirmed theory which effects an economical

systematisation, both deductive and inductive, of a large class of particular facts and empirical generalisations, and which is heuristically fertile in suggesting further questions and new hypotheses. And as far as suitability for individual systematisation, along with economy and heuristic fertility, are considered essential characteristics of a scientific theory, theoretical terms cannot be replaced without serious loss of formulations in terms of observables only: the theoretician's dilemma, whose conclusion asserts the contrary, starts with a false premise.

I agree — we need "non-observable" terms, or "theoretical" terms, but I say this because I am thinking of these terms as linguistic terms, i.e. as uninterpreted terms of some suitable first-order language in terms of this text. We can express our knowledge of science only by abstracting certain features of reality and expressing our knowledge about these features in some language. Applying this scientific knowledge asks of us to interpret our scientific language, and this enables us to determine whether the terms in our language refer ultimately, after many complex twists and turns, to some objects or relations in some real system. Lewis (1970, p.428) remarks after all that a term that may be defined by means of other terms that may be shown to have sense and denotation can "... scarcely be regarded as a mere bead on a formal abacus".

5. RAMSEY SENTENCES AND PUTNAM'S PARADOX

In conclusion of this chapter, let us look again at the question that led Ramsey to define his particular kind of "Ramsey sentences". He wanted to find a way to determine the so-called "empirical content" of a theory. Or, in other words, the question is as Carnap (1966, p.248) phrases it,

How can theoretical terms, which must in some way be connected with the actual world and subject to empirical testing, be distinguished from those metaphysical terms so often encountered in traditional philosophy — terms that have no empirical meaning?

Ramsey proposed that the entire set of theoretical and correspondence postulates of a theory be replaced by a certain kind of sentence in which theoretical terms do not occur at all. Suppose we consider theory T with n theoretical terms, $T_1, T_2, \ldots T_n$ which are introduced by the theory's theoretical postulates. Now say these theoretical terms are connected to the empirical world by the theory's c-rules which are such that they contain m observational terms, $O_1, O_2, \ldots O_m$. We know that the positivists take a theory to be a conjunction of theoretical postulates and all the c-rules (called "correspondence postulates" by Carnap (ibid., pp.248ff.)). Thus if we want to formulate the theory fully we shall have to combine the set of T-terms with the set of O-terms such that we get $T_1, T_2, \ldots T_n, O_1, O_2, \ldots O_m$. Now what Ramsey did was to propose that we substitute all the theoretical terms in this set with a set of so-called "correspondence variables" (ibid.): $U_1, U_2, \ldots U_n$, and then add existential quantifiers — $(\exists u_i)$ prefixed — to this formula (to bind the variables), in order to form what has become known as a "Ramsey sentence". (There are many explications of Ramsey sentences. See for instance Carnap (ibid.) for an easy step-by-step explication, or Ramsey's own article on the topic in Braithwaite's *The fundamentals of mathematics* (1931), or Lewis (1970). See also Psillos (1999b).)

This kind of existentially quantified sentence asserts that there exists at least one entity of the kind that will satisfy the relevant sentence. Note though that the variables U_n with which we have substituted the n number of theoretical terms do not refer to any *particular* entity or set (or class) of entities, given the meaning of the existential quantifier. Carnap (ibid., pp.251, 252) writes: "The assertion is only that there is at least one class that satisfy certain conditions. The meaning of the Ramsey sentence is not changed in any way if the variables are arbitrarily changed". Also note that any empirically verifiable sentence that can be derived from the theory can also be derived from the Ramsey sentence.

Still, one might wonder, so what? Well, the important thing about Ramsey sentences to Carnap at least was that "... we can now avoid the troublesome metaphysical questions that plague the original formulation of theories" (ibid., p.252). How? Carnap's (ibid.) answer is that a Ramsey sentence

> ... continues to assert, through its existential quantifiers, that there is something in the external world that has all those properties that physicists assign to the electron. It does not question the existence — the 'reality' — for this something. It merely proposes a different way of talking about this something. The troublesome question it avoids is not 'Do electrons exist'? But, 'What is the exact *meaning* of 'electron'', because the term itself does not appear in Ramsey's language.

Having said this, notice also (Carnap's (ibid.) point) that Ramsey did not in any way imply that a theory in its Ramsey-sentence-form is a sentence in an ordinary observation language. What Ramsey's sentences require is an *extended* observation language, which is a language that is observational because it contains no theoretical terms, and "... which has been extended to include an advanced, complicated logic, embracing virtually the whole of mathematics" (ibid.).

From a model-theoretic point of view there is little to choose between the following two scenarios, and we can effortlessly move between them in a one-to-one way:

- Scenario One: We have T_i in the (theoretical) vocabulary of the language and choose some <u>interpretation</u> of T_i in the domain D. ("Choose" here is not meant either constructively or ostensively, but rather in an ordinary common mathematical sense.)

- Scenario Two: T_i is not in the vocabulary, but we choose some <u>assignment</u> to variable U_i of an appropriate object in D (which may also denote subsets or relations of D) to realise $\exists U_i$.

Assignments are denotation functions (interpretations); assignments simply assign values from D to variables, where interpretations do so to predicate symbols or individual constants. Think, analogously, of the different, but equivalent, axiomatisations of the mathematical notion of "group" in languages with different vocabularies.

Returning to Carnap, he (ibid., p.254) reminds us that the theoretical language somehow seems to presuppose that theoretical terms refer to something that is somehow *more* than what the context of the theory offers us. He (ibid.) writes:

Some writers have called this the 'surplus meaning' of a term. When this surplus meaning is taken into account, the two languages are certainly not equivalent. The Ramsey sentence represents the full *observational content* of the theory. It was Ramsey's great insight that this observational content is all that is needed for the theory to function as a theory, that is, to explain known facts and predict new ones.

According to Hempel (in Feigl et al. (1958, pp.76 ff.)) though, Ramsey's method of "transforming" theoretical sentences into observational ones is rather ineffective. On the so-called "Ramsey method", Hempel (ibid., p.81) remarks that a Ramsey sentence

... still asserts the existence of certain entities of a kind postulated by T' [the interpreted theory], without guaranteeing any more than does T' that those entities are observable or at least fully characterisable in terms of observables. Hence, Ramsey-sentences provide no satisfactory way of avoiding theoretical concepts. And indeed, Ramsey himself made no such claim. Rather, his construal of theoretical terms as existentially quantified variables appears to have been motivated by considerations of the following kind: If theoretical terms are treated as constants which are not fully defined in terms of antecedently understood observational terms, then the sentences that can formally be constructed out of them do not have the character of assertions with fully specified meanings, which can be significantly held to be either true or false; rather, their status is comparable to that of sentential functions, with the theoretical terms playing the role of variables. But of a theory, we want to be able to predicate truth or falsity, and the construal of theoretical terms as existential quantified variables yields a formulation which meets this requirement and at the same time retains all the intended empirical implications of the theory.

Hempel's last point above is also discussed — and its problematic nature perhaps made more clear — in the context of a paper by Stathis Psillos entitled "How not to structure realism", read at the LMPS conference in Cracow in 1999 (part of which is taken up in Psillos (1999b)). Psillos addresses what he refers to as Carnap's "neutralism" towards the realist/instrumentalist divide (recall Carnap's words quoted at the beginning of Section 3.1 above) in Chapter 3 of his *Scientific realism: How science tracks truth* (1999b). I shall in what follows comment first on Psillos's remarks on Carnap and then, in conclusion of this (rather long) section turn towards the issues implied by Putnam's model-theoretic paradox.

Psillos (ibid.) reminds us that Carnap (see also Carnap (1956)) studies the structure of scientific theories in terms of two (sub-)languages. We have a language L_O, where the variables of this language denote "physical things" and range over a finite domain. We also have a language L_T, where V_T is the vocabulary describing T-terms, and where included into L_T we find a type-theoretic logic with an infinite sequence of domains $D_0, D_1, ...$, where D_n is the nth level domain and D_{n+1} is the powerset (i.e. the set of all subsets) of D_n. Each variable and individual constant belong to one of these domains, and this, together with the fact that D (the union of all the D_i's) is taken as the domain of classical mathematics, imply that the variables of L_T range over mathematical entities. Psillos (ibid.) points out that in these terms it is taken that the elements of D can be shown to represent all the "physical concepts" in theories. In this sense L_T may be shown, for example, to accommodate a space-time co-ordinate system such that a four-tuple of numbers represents each space-time point. In this way space-time magnitudes are (represented by) functions from these four-tuples of numbers (i.e. space-time points) to numbers. Physical objects in these terms are depicted as four-dimensional regions inside which relations of physical magnitude have a certain distribution.(Ibid.)

In this context Carnap defined two senses of "real". If it is stated in L_O that some observable event is "real", we may conclude that the sentence in this language describing this event is, indeed, "true". The second sense of "real" is given in the more complex theoretical case. Carnap (1956, p.45) writes:

> For an observer to 'accept' the postulates of T, means here not simply to take T as an uninterpreted calculus, but to use T together with special rules of correspondence C for guiding his [sic] expectations by deriving predictions about future observable events from observed events with the help of T and C.

Carnap is referring here to the so-called "excess empirical content" or "surplus content" of T-terms (briefly already referred to in the previous section). Psillos (ibid.) notes that stating the latter does not imply stating that T-terms have factual content, although it does imply acknowledging that L_T is more than a mere syntactical "construct" or tool.

As we know, Carnap takes D as the domain over which the variables of the theoretical language range, and, according to him, the structure (power hierarchy) of the domain D is isomorphic to the structure built on the set of natural numbers. In Carnap's view, that is all that we can say about a certain theory's domain. We can specify its structure uniquely, but we cannot say anything specific about the elements of the structure. This is why Psillos refers to Carnap (in a certain sense) as a structural realist. Carnap simply asserts that a theory T exemplifies a certain mathematical structure connected to reality by C-postulates. And if he does say that the kinds of entities designated by the theoretical assertions of T are mathematical entities (i.e. natural numbers or classes, etc.) he is not falling into any metaphysical traps anyway, since he still escapes saying anything explicit about the nature of the elements of the reality that T is "about".

Now obviously if we accept that T-terms are somehow "real", we are overstepping the boundaries of empiricism in a very basic way. We know also that Carnap did not want to concern himself with answering ontological questions in their traditional metaphysical sense. Psillos (ibid.) refers us to Carnap's distinction between internal and external questions — set out in Carnap's (1950) "Empiricism, semantics, and ontology" — to remind us how Carnap proposed to eliminate the metaphysical implications of pondering the surplus content of T-terms. External questions are branded metaphysical, because they imply answering questions of reference to reality in terms of the whole of reality, i.e. without reference to some discourse or other. Questions of this kind thus answers questions of reference of certain entities after already having established or acknowledged the existence of these entities. Internal questions, on the other hand, are answered from within some linguistic framework which enables us to talk about the entities at issue. This implies that if we accept the relevant framework, we accept (arguments made from within the framework for) the existence of such entities.[4] So, all that Carnap had to do to escape traditional metaphysics was to classify realist questions concerning theoretical terms as internal such that statements about the existence of particular entities are taken as empirical assertions following from some accepted linguistic framework.

Returning to Ramsey sentences, Psillos (ibid.) notes that such sentences enable us to use theoretical terms and predicates to examine the empirical content of theories, without having to think of these terms as "names" for anything. Formulating the Ramsey sentence of a theory presents us with a sentence which has a truth value and which can express a "judgement". Recall that where TC is the theory and is represented by $TC(t_1, \ldots t_n, o_1, \ldots o_m)$, where TC is any n+m predicate, the Ramsey sentence is given by $\exists u_1, \ldots \exists u_n TC(u_1, \ldots u_n, o_1, \ldots o_m)$ (ibid.). So the Ramsey sentence $\exists u TC(u,o)$ implies the empirical (or factual) content of T, but it also contains an abstract claim of realisation, because it implies that (at least some) statements of the form "u stands in relation TC to o" are true. And thus (ibid.) it is better to use the weaker $\exists u TC(u,o)$ (rather than TC(t,o)), since then we can use T and understand what it "says" "about" reality, without specifying information concerning the particular classes which realise $\exists u TC(u,o)$. Thus, as Psillos (ibid.) notes, two parties may disagree over what the exact realisation of a given Ramsey sentence is, but still use the same Ramsey sentence to derive the observational consequences of the theory.

Braithwaite (1953, pp.80, 81) comments that this implies a compromise between the realist claim that a theory is true (i.e. that some theoretical terms exist) and the empiricist claim that accepts that T-terms may be real, without saying that the T-terms of some theory have factual reference. He (ibid., p.79, also quoted by Psillos) describes the Ramsey-status of a theoretical concept, for example "electron" as follows:

> There is a property E (called 'being an electron') which is such that certain higher-level propositions about this property are true, and from these higher-level propositions there follow certain lowest-level propositions which are empirically testable. Nothing is asserted about the 'nature' of this property E; all that is asserted is that the property E exists; i.e. that there are instances of E, namely electrons.

Now, as we have seen, Carnap seems to think of the classes of theoretical entities related to observational entities by the relations in the original theory as classes of mathematical objects since he posits (Psillos's point) an extensionally identical mathematical function corresponding to each descriptive theoretical constant of the theoretical language such that mathematical entities designated by such functions are extensions of these descriptive constants. In other words it is classes of mathematical entities that realise a Ramsey sentence. Psillos's point is that Carnap's correlation of mathematical entities with observable phenomena implies that the Ramsey sentence of a theory "preserves the structure" of the theory.[5] Thus, accepting a certain Ramsey sentence implies commitment to the observational consequences of TC and to a certain "logico-mathematical" structure in which (a description of) observational phenomena is embedded (ibid.).[6] This also implies commitment to certain abstract existential claims such that there are non-empty classes of entities that realise the structure.

We shall not analyse issues concerning analyticity here, but we have to consider briefly the special kind of sentence Carnap proposed in this context in terms of theoretical sentences, to clarify issues concerning the realisations of Ramsey sentences (if any) within a realist context. In order to avoid T-terms, Ramsey proposed to use a particular Ramsey sentence instead of the postulate TC. To address analyticity in terms of theoretical languages, Carnap (ibid.) retains the T-terms, but specify different postulates for distinguishing between analytic and synthetic content. He takes the

Ramsey sentence as a P-postulate, that is, as a physical (synthetic) postulate. Carnap (in Hintikka (1975, p.82)) claims that the concept of analytical truth can be "... defined on the basis of meaning postulates which represent meaning relationships which hold among the primary descriptive constants", and he refers to these postulates as "A-postulates".[7] Carnap formulates as A-postulate A_T for the T-terms, the conditional sentence R ⊃ TC (or in Psillos's terms, ∃uTC(u,o) ⊃ TC(t,o)). This sentence is purely analytic, because its (semantic) truth rests on the meanings of the relevant theoretical terms. This sentence together with the Ramsey sentence itself, that is [R & (R ⊃ TC)] then logically imply the theory (Carnap, 1966, p.270).[8]

Now, bringing us back to Hempel's (1958, p.81) point that since we want to be able to determine the truth or falsity of theories, the Ramsey construal of theoretical terms as existential quantified variables "... yields a formulation which meets this requirement and at the same time retains all the intended empirical implications of the theory", Psillos (1999a, p.1) writes:

> ... there is a big asymmetry between an *interpreted scientific theory* and its *Ramsey-sentence*. Whether an interpreted theory is true or false is an empirical matter. In particular, it's possible that a theory can be empirically adequate and yet false. However, if the Ramsey-sentence of a theory is empirically adequate at all, it is *guaranteed* to be true, i.e., *without further constraints*, it is guaranteed that *there is* an interpretation of the second-order variables which makes the Ramsey-sentence true.

In model-theoretic realist terms, given the non-monotonic analysis of over-determination offered in Chapter 2, a theory can indeed be empirically adequate, but turn out later to be "false", but only "false- in-a-certain-model", of course. This may happen if the model in which the theory had been interpreted later proves to be inadequate, but the relevant empirical reduct into which some empirical model were embedded turns out to be an empirical reduct of a different or "new" model of the theory, still perhaps with the same isomorphic relation with the same empirical model. (See Chapter 2.) A Ramsey sentence however is merely a sentence in some empirical reduct, in my terms, so that should such a sentence be "empirically adequate" this implies that one of my relations of isomophism between the relevant empirical reduct and some empirical model holds.

Why does Psillos though claim that Carnap's account of Ramsey sentences imply that any empirically adequate Ramsey sentence is true? Well, Psillos (ibid.) states that if the domain of a theory is viewed merely as a set of objects which has no natural structure, the domain can always be so "carved up" that the Ramsey sentence of the theory is true of it. In other words, the antecedent of a Carnap sentence ∃uTC(u,o) ⊃ TC(t,o) will always be true without empirical testing. Even worse, is that TC(t,o) will also always be true, if the Ramsey sentence is true and the Carnap sentence is given (by good old *modus ponens*).

Now Psillos (ibid., p.4) writes that in Carnap's terms a realist accepts both the Ramsey sentence, ∃uTC(u,o), and the meaning postulate, ∃uTC(u,o) ⊃ TC(t,o); while a defender of Ramsey-sentences merely takes the Ramsey sentence as acceptable. Psillos (ibid.) points out though that ∃uTC(u,o) ⊃ TC(t,o) has no more empirical content than the Ramsey sentence, and so that Carnap did not think that a realist takes "extra empirical risks" (ibid.) — i.e. more than a Ramsey defender — by asserting the

THE STATEMENT ACCOUNT OF SCIENCE 83

existence of electrons for instance. And, moreover, an advocate of Ramsey sentences could come to accept that *if* the Ramsey sentence is true, then there are electrons. This would imply that she accepts a meaning postulate, but does not necessarily imply that she goes "beyond the limits of empirical inquiry" (ibid.). Psillos (ibid.) concludes from this that structural realism is thus a compromise between "enlightened empiricism" (i.e. empiricism that accepts the existence of theoretical entities) and realism.
He writes:

> ... all we need to do in order to effect the compromise is adopt a meaning postulate ($\exists u$ $TC(u,o) \to TC(t,o)$) for an n-tuple of T-terms and an m-tuple of O-terms, which says: *if* the world is so constructed that there are classes of entities which satisfy $R_{(TC)}$ [i.e. the Ramsey sentence], then the T-terms are to be understood in such a way that they designate these classes. (Ibid., p.5)

Obviously, we have to leave it "open" or undecided whether the antecedent of this conditional is true or false, and this will imply that it is undecided whether the T-terms denote anything or not — if the Ramsey sentence is true, the n-tuple of T-terms of the theory does designate something, while if the Ramsey sentence is false, it obviously does not. So, as Psillos (ibid.) points out, if no external constraints or limitations are put on the Ramsey sentence, it will always be true, and then by *modus ponens* $TC(t,o)$ will always be true, which will imply that no empirical investigation is necessary to discover what the relevant unobservable entities are. Note that to a defender of model-theoretic realism, stating that the Ramsey sentence will be "true" or "false" does not make sense, since model-theoretically, a sentence can only be true or false in some *interpretation* (or model). Anyway, Psillos does not want to believe in this trivial way in which to "make" Ramsey sentences come out "true", and finds a way out in what we can call the Russell-Newman problem, following Psillos (ibid.).

If we consider the structure, call it W, generated by the relation TC (recall that TC is the theory and is represented by $TC(t_1, ... t_n, o_1, ... o_m)$, where TC is any n+m predicate, and the Ramsey sentence then is given by $\exists u_1, ... \exists u_n TC(u_1, ... u_n, o_1, ... o_m)$) we find that this structure is a set of ordered tuples of the individuals of this domain. In this context Psillos (ibid., p.2) writes:

> Suppose, as the structuralist would have it, that we posit the existence of a set of theoretical entities. Can this set *fail* to possess the required structure W, whatever that be? The answer should be clearly negative. For the domain — considered as a set — possesses *all* structures which are consistent with the number of its elements. Intuitively ... if all we aim is to show that there is *some* relation which generates structure W, there is nothing to stop us from arranging the elements of the domain in tuples such that they correspond to the required structure W. (Ibid.)

Psillos (ibid.) bases this on the formal feature of set theory that every set E determines a so-called "full structure", that is one which contains all subsets of E, and so every relation-in-extension of E. And so, since "... all relations-in-extension are contained in the posited domain of unobservable entities (considered as a set), it follows that one can never fail to generate the required structure W on this domain" (ibid.). And so to claim that there is a relation such that the structure of the unobservable world is W, is to claim almost nothing important. All that is actually claimed is that the assumed domain of unobservables must have a certain cardinality (ibid.). This Psillos refers to as the "Newman challenge" or the problem of "trivial realisation". Newman proposed that we rather talk only of the structure of a set of individuals if some relations are specified. Well, which relation, or set of relations, can we choose to "structure the domain of unobservables"? The structuralist answer, according to Psillos (1999b) is a non-answer, since they find their assertion that there is some relation TC, and all that is known of it is the structure W that it generates, sufficient.

There is also the problem of multiple realisations of Ramsey sentences to take into account, also mentioned by Lewis (1970). Lewis (ibid., p.430) defines a realisation of a theory T as follows. Let us write the postulate of T in terms of its T-terms as $T[t_1, t_2, \ldots t_n]$. Then Lewis (ibid.) states that if we replace the T-terms uniformly with variables $x_1, x_2, \ldots x_n$ respectively, where these variables are distinct and do not occur in the postulate already, we get a so-called "realisation formula" for T: $T[x_1, x_2, \ldots x_n]$.[9] In this context then, notice (ibid.) that a Ramsey sentence of a theory T states merely that T is realised, while a so-called Carnap sentence of T (i.e. the sentence $\exists u \, TC(u,o) \to TC(t,o)$ as defined above) is neutral as to whether T is realised or not. The Carnap sentence states merely that *if* the Ramsey sentence is true, the theoretical terms should be understood such that the theory is true. Does a Carnap sentence specify interpretations of all T-terms though? Lewis (ibid.) points out that there are three cases to be considered:

- If T has a unique realisation: the Carnap sentence does give the correct specification in the sense that t_1 names the first component of a unique realisation of T, t_2 names the second component, and so on. (We shall discuss this option below, just before we move on to Putnam's paradox.)

- If T is not realised: the Carnap sentence says nothing about the denotation of T-terms in this case, because T-terms of uninterpreted theories do not name anything.

- If T is multiply realised: the Carnap sentence tells us that the T-terms name the components of some realisation or other, but it does not tell us which one, and there seems to be no unique way of choosing one. This happens in the case of structural realism since structure determines its domain only up to isomorphism, and so many different domains may realise the same structure.

This seems very instrumentalist, since it implies that either the T-terms in a Carnap sentence do not name anything, or they name the components of a non-unique realisation of T. Model-theoretically, and also according to Psillos, this is not a real problem, since we agree that there are ways to distinguish between important and unimportant interpretations of a domain.

Returning to our problem of trivial realisation though, not even fixing or determining the intended interpretations of the domain helps here. We have to specify the nature of the relation TC in which there are entities (*t*) standing to *o*. And Psillos (ibid.) points out that Carnap's case is even worse, since he claims that all that the theory has to assert is that there are mathematical entities which stand in relation TC (taken as a "logico-mathematical predicate") to the observed phenomena. Moreover the domain of L_T is the type hierarchy D based on the set D_0 of natural numbers, and so contains all classical mathematical structures. So, this domain can represent any structure and therefore also the observational structure imposed by TC. The richness of the domain guarantees the truth of $\exists u\ TC(u,o)$ in Carnap's case.

Carnap seems untroubled by the fact that as far as his view can be portrayed as structural realism he cannot get out of this, since he cannot restrict the relations defined on a certain domain within his account. Psillos (ibid.), on the other hand, points out that if we try to restrict the relations to those that are "real", we are moving beyond structure, since this implies that we know somehow which extensions are "natural", i.e., which subsets of the power set of the domain corresponds to "natural properties and relations" (ibid.). This seems like the only option to rescue us from Newman's challenge though. Psillos (ibid.) comments that the only way in which we can introduce this kind of "non-structural knowledge" is to use the information concerning which properties and relations are the "natural constituents" of the world, which is offered to us by interpreted scientific theories. For reasons of space I cannot discuss the latter comment here, but shall rather now briefly introduce my view of Newman's challenge.

A model-theoretic realist may be able to escape this challenge. On the one hand, as mentioned already, model-theoretically, given that a Ramsey sentence contains no theoretical terms, it may be a sentence interpretable in some empirical reduct of some interpretative model of a theory. Terms in empirical reducts can only be said to refer to some system in reality if it is possible to find some empirical model isomorphically embedded into the relevant reduct. Such a relation of isomorphism implies that the (Ramsey) sentence in the given reduct is "true", because it refers, i.e. because the theory is found to be empirically adequate, because *the world* is such that it makes this possible. The world exists independently of us, and this relation of isomorphism must be *found* to hold, *proved* to hold, and this cannot be done in an *a priori* way.

On the other hand, equally, or perhaps more importantly, it is *not only the world* that restricts us, but also *the theory* itself. An empirical reduct is a structure obtained by ignoring some of the relations in a model, and recall that a model is a mathematical structure in which the relevant theory *is true*. Our "choices" (of models of a certain theory) made to show empirical adequacy, are no choices in the true sense of the word. Rather they are options determined — or restricted — *both* by our theory (being true in the models we choose, and so in their empirical reducts) and by the real world (i.e. the fact that some empirical model — conceptualising data about the relations and objects of some real system — must be found to be isomorphically embedded into — wait, not just any old empirical reduct — but into one in which the relevant theory is indeed true). Also remember that the theory links together all the relations, theoretical and observational, in many ways. More about this when we discuss Putnam's paradox below.

Recall that Lewis (1970) claims that T-terms refer only if there exists a unique realisation of the theory. This in itself also still faces Newman's challenge. However, Psillos (ibid.) proposes that we view uniqueness of this kind to imply the requirement that there is a unique relation-in-extension TC which structures the specific domain such that the entities of this domain stand in TC to o. But, this again requires that the domain is already "carved up" in natural kinds with natural relations holding among them.

It seems then that only if we allow this kind of knowledge about the real system (or domain in the above terms) can we escape the construction of trivial extensions generating the required structure, since in such a case our extensions will have to be such that they do not violate this knowledge we have of the domain, and "getting" only these extensions is not a trivial case at all. So, again, model-theoretical realism offers us also such an "external standard" to "compare carvings of the domain" (ibid.) in the form of the relation of isomorphism of some empirical model (i.e., from the side of reality, as it were) into some reduct (of some conceptual, interpretative, mathematical model) in which the theory is true (now, from "the side of the theory").

So, Psillos (ibid.) portrays Carnap's account of scientific theories as moving from strict empiricism via the "Ramsey way" to accommodate Carnap's "neutralism". Then to retain the latter, Carnap turns towards structural realism, only to find that structural realism can actually only be meaningful if it allows something "more" than simply asserting that the structure of the unobservable world can be known, which might just serve to turn Carnap (albeit after the fact perhaps) into a scientific realist. The kind of "metaphysical thesis" that Carnap would have to accept — i.e. the ontological claim that reality exists independently of us, and that reality "is" in a certain way independent of our knowledge claims — is, at least for me, simultaneously the precondition of my realism, and all the "metaphysics" I allow for my realist argument. More about this in Chapter 5.

This brings us to the last issue to be discussed in this section of this chapter, namely that of Putnam's paradox. Briefly, in very simple model-theoretic terms (since we need nothing more elaborate to state the paradox and show its weakness) Putnam's (1978) argument comes down to the following: Take T_1 to be an "ideal" theory in all possible senses, except being objectively true, (i.e. it is left "open" whether the theory is "objectively" — whatever that may mean! — true or false). Take as our domain of reference (of interpretation) parts of "THE WORLD", so that we have "things" (objects) that may be referents of singular terms and classes of "things" that may serve as referents of general terms. Assume that "THE WORLD" can be divided into infinitely many pieces. Take T_1 to claim that there are infinitely many things (such that, in this respect, T_1 is objectively correct about the world). Then, by the completeness theorem, and given that T_1 is consistent (i.e. is satisfied by some model) and has only infinite models, T_1 has a model of every infinite cardinality.

And then, if we choose a model M of the same cardinality as "THE WORLD" and we map the individuals of M one-to-one into the pieces of "THE WORLD", and we then use this mapping to define relations of M directly onto "THE WORLD", the result is a satisfaction relation "SAT" which is a "correspondence" between the terms of the language L and the sets of parts of "THE WORLD" such that T_1 comes out true of "THE WORLD" — provided that "true" is interpreted as meaning model-theoretic truth, i.e. "TRUE(SAT)". Translated into model-theoretic terms, in other words, we take an interpretation as a model-theoretic function defined over a certain domain and assigning specific denotations to the non-logical predicates and individual constants of the language, and an interpretation is a model of the language if the sentences of the language are satisfied ("comes out true") under the interpretation, i.e. if the mappings of the elements of the language provided by the interpretations are in the denotations of its (the language's) predicates (which are subsets of the domain of discourse).

Thus, as Lewis (1984, p.221) points out, what Putnam implies is that we cannot imagine in a sensible way that an ideal empirical theory can be false because the world does not turn out to be the way it says it is. Putnam implies that since there seems to be no definite way of linking our words with their referents, reference can be construed any way we want, except for the fact that

> ... there is one force constraining reference, and that is our intention to refer in such a way that we come out right; and there is no counterveiling force; and the world, no matter what it is like (almost) will afford *some* scheme of reference that makes us come out right; so how can we fail to be right? (Ibid.)

The paradoxical implication of Putnam's argument is that *if* the criteria we formulate to determine the truth of certain sentences are indeed *solely internal* (implying that these sentences "must come out true" (Van Fraassen, 1997, p.18)), then "practically any theory is true" (ibid.). I.e. the dilemma is — or seems to be — that if we cannot talk about "THE WORLD" then we cannot say anything about the extensions to our predicates in "THE WORLD"; but if we can indeed describe — talk about — elements in "THE WORLD", then we can distinguish between right and wrong assignments of extensions to our predicates in a rather trivial manner. (See Putnam (1978), Lewis (1984), and Van Fraassen (1997) for more detailed definition and discussion of this argument.)

Due to limitations of space I shall not analyse in detail Putnam's claims (Putnam, 1983, p.2, 3) concerning the role he ascribes to the Löwenheim-Skolem theorem in stating his paradox[10], save for the following (and noting that he assumes the ideal theory can be formulated in a first-order language such that the Löwenheim-Skolem theorem can hold). I agree with Lewis (1984, p.229) that it is not this theorem that should get "star-billing" in Putnam's argument. Putnam (1983, p.3) writes:

> Now the argument Skolem gave, and that shows that the 'intuitive notion of a set' [i.e. the so-called intended interpretation] ... is not 'captured' by any formal system, shows that even a *formalisation of total science* (if one could construct such a thing), or even a *formalisation of all our beliefs* (whether they count as 'science' or not), could not rule out denumerable interpretations, and, *a fortiori*, such a formalisation could not rule out *unintended* interpretations of this notion.

In Putnam's context, Lewis (1984, pp.229, 230) writes:

> In fact, what is needed is pretty trivial. ... (almost) any world can satisfy (almost) any theory. The first 'almost' means 'unless the world has too few things'; the second means 'unless the theory is inconsistent'. This premise is obtained as follows. A consistent theory is, by definition, one satisfied by some model; an isomorphic image of a model satisfies the same theories as the original model; to provide the makings of an isomorphic image of any given model, a domain need only be large enough. The real model theory adds only a couple of footnotes ... . First, by the Completeness Theorem, we could if we wished redefine 'consistent' in syntactic terms. Second, by the Löwenhiem-Skolem Theorem, our 'unless the world has too few things' is less of a qualification than might have been

supposed: any infinite size is big enough. But the qualification wasn't important in the first place. If Putnam's thesis had been that an ideal theory can misdescribe the world only by getting its size wrong, that would have been incredible enough. And in fact that *is* Putnam's thesis: for all he said, it is still possible for an ideal theory to misdescribe a finite world as infinite. Who cares whether the possibility of similar mistakes among the infinite sizes also is granted? We thought it was possible to misdescribe the world in ways that have nothing to do with its size.

Van Fraassen (1997, p.18) states that what we must do, is find external constraints "... binding or glueing our words to things quite independently of our intentions and desiderata, and giving factual intent to our theorising" (ibid.). Model-theoretically speaking it is part of the basic conditions of science (here in contrast to mathematics and logic) that these criteria or constraints are not solely internal. A "successful" mapping between the individuals of an empirical model embedded into an empirical reduct of some interpretative model of a given theory and the entities and relations of some real system depends just as heavily on the nature of the real system as it does on the logical relations of satisfaction between the theory, its interpretative model(s), and its empirical model(s). (See *Figure 3* in the next chapter, to see schematically the "two-direction" aspect of model-theoretic realism at issue here.)

Van Fraassen (ibid., p.20) points out the trivial fact that not only does one mapping between the ideal theory and "THE WORLD" exist in Putnam's ideal case, but indeed obviously many such mappings exist. The actual question thus is: how do we know which one to pick? Well, again, we do not do anything as trivial as merely "picking". A "successful" mapping is determined *both* by our scientific activities that gel finally into a certain empirical model of a given theory, and by the real system of which the empirical model in question is offering a certain (scientific) image.

If we consider *Figure 3* illustrating model-theoretic realism given in the next chapter (Section 4.6), we shall see that model-theoretic "mappings" are taken to come from "two directions". The relations between a theory and its models and the formulation of an empirical reduct of some model may be done by us, as it were, based on our knowledge and our application objectives. The formulation of empirical models and, consequently, the formulation of the isomorphic relations between empirical models and empirical reducts, however, are determined by the way in which the world is, and we do *not* determine that. (See Chapter 5.) So, we cannot *choose* a model M of the same cardinality as "THE WORLD", as Putnam proposes, rather we *find* that such models exist, here after the fact, as it were.

So, in the final analysis, in the model-theoretic realist case we, or rather our interpretations, are ultimately constrained by the way the world is — *independent* of what we theorise about it. In the context of empirical models being conceptualisations of experimental results, let us briefly consider the problem of transduction[11], which refers to the problem of "carrying over" the results of some idealised (experimental in this case) context to the complexities of real systems. Bhaskar (1986) stresses the fact that scientists produce the *empirical grounds* for the laws of nature in their laboratories, but not the *laws themselves*. (In my terms this refers to the distinction between science, the "reality" of science, and "Nature" that I discuss in the first sections of Chapter 5.) Bhaskar (ibid., p.30) claims that distinguishing between "real and universal ... but non-empirical laws and their real and empirical but contextually localised grounds" dissolves the problem of transduction. The justification for each individual law can thus be found in the Bhaskarian stratification of nature, and not by tying laws to closed systems and *ceteris paribus* conditions (as "classical" empiricists seem to do).

Bhaskar also claims that causal laws are "ontologically uncoupled" (Bhaskar 1986:44) from patterns or sequents of events. Empirical regularities only occur as a result of active *interference* in nature: therefore this ontological distinction — between the empirical regularity that scientists produce (in the transitive dimension) and the causal law (in the intransitive dimension) that it enables us to identify — has to be presupposed and acknowledged if experimental (and thus scientific) activities are to be comprehensible. If something like Bhaskar's transcendental realist assumptions about the relations between reality and science are not made (which can be done without buying into his metaphysical notion of the tendencies of nature), any scientist could generate any pattern of events at will, rendering all scientific activity totally uninteresting. In other words, a realist philosophy of science should indeed be able to sustain an epistemological relativity, but fight against surrendering to ontological relativism — model-theoretically we fight this with the assumption of the existence of an independent reality (discussed more fully in Chapter 5).

In line with the above is Lewis's (1984, p.226) remark that referring is not just something we *do* — "What we say and think not only doesn't settle what we refer to; it doesn't even settle the prior question of *how* it is to be settled what we refer to. Meanings — as the saying goes — just ain't in the head".

In the context of our discussion concerning Putnam's claims, we note that Van Fraassen (1997, p.36) explains the problem concerning the "fixing" of the reference of language terms as follows:

> ... each of our predicates has an extension, and might have had a different extension. But unless they have the right extension, we can't use our language to frame genuine, non-trivial empirical statements or theories. So, under what conditions do they have, or acquire, the right extensions? (Ibid.)

Later on he (ibid.:37) comments that we obviously do not do any kind of fixing and whether Nature does any kind of "glueing" is not really a sensible question — in any absolute sense I might add. Model-theoretically speaking, language has to be interpreted nominalistically (in the qualified sense of nominalism discussed in Chapter 5) so that science can be interpreted realistically. Only if we can handle the fact that there are no fixed one-to-one unique mappings between universal language terms and objects in reality, do we have a chance of making sense of the unfixability of reference. Paradoxical as this may sound, this is the only non-metaphysical way to be a scientific realist.

So, again I agree with Lewis (1984, p.231) when he writes that even if Putnam's model-theoretic argument did not fail, it would not mean the end of realism. Lewis points out that there would still be a world independent of us, it would still have many parts which may fall into classes and relations, there would still be interpretations and ascriptions of reference (intended and unintended ones), truth of a theory would still make sense, even truth under all intended interpretations. Lewis (ibid.) writes that the only "trouble" Putnam really points out is the fact that truth under intended interpretations comes too easily because there are so many intended interpretations, and acknowledges that this is "trouble", but not necessarily "anti-realist trouble". The application of minimal model semantics, added to my model-theoretic analysis of the language of science (Chapter 2), proves that there still is a way in which to trace reference relations to reality even in the face of "too many intended interpretations", or, as I refer to it, in the face of "empirical proliferation". So, realism *can* be rescued.

Let us now look at some of the main non-statement accounts — including Nancy Cartwright's "simulacrum" account — of science to clarify the main points of a model-theoretic realism even further.

NOTES: CHAPTER 3

[1] As we shall see in Chapter 4, the model-theoretic view on this is that meta-mathematics *is* model theory, which, in a sense *is* set theoretical, so that the statement non-statement distinction becomes more one of emphasis, than of real differences.

[2] Remember that Cartwright takes scientific theories to be a set of fundamental laws — like Maxwell's equations — from which explanations in physics are supposed to start. See Chapter 4.

[3] This is one more reason for the importance of theories or fundamental laws in the process of science. One theory may have infinitely many "incompatible" models (i.e. models that are *not* isomorphic) *and* "the facts" may be valid in all of these.

[4] Recall that for Carnap the choice of such frameworks is a pragmatic matter. Also important is that Maxwell (1962) demanded that a linguistic framework should be broad enough to allow for the development of explanations of phenomena in terms of novel empirical predictions.

[5] Psillos (ibid.) notes that Carnap had to posit this identity between theoretical and mathematical concepts, because he believed at all costs that a linguistic framework for analysing the language of science must be broad enough to include future theoretical concepts. Given this identity, new physical concepts can be taken up in this framework, since we can provide the relevant extensionally identified mathematical functions, since the extensional language L_T is such that any physical magnitude of any logical type will be identifiable with a certain mathematical function. This also allows for comparison of theories in the sense of finding mathematical functions corresponding to different theoretical concepts and examining whether these functions are extensionally identical.

[6] See Psillos's (ibid.) comments in terms of Carnap's method of extension and intension on Feigl's criticism of Carnap's "instrumentalist" interpretation of Ramsey sentences in the sense that it seems that Carnap takes theories to be mathematical models with observable phenomena embedded in them.

[7] Note that A-postulates do not tell us anything about the physical world, but merely specify the meaning relations holding between the descriptive terms of our language.

[8] So this sentence A_T which contains all T-terms, together with the Ramsey sentence (or F_T) which contains no T-terms, say just as much as the postulate TC.

[9] "Any n-tuple of entities that satisfies this formula, under the fixed standard interpretations of its O-terms, may be said to *realise*, or to be a *realisation* of the theory T" (ibid.).

[10] The Löwenheim-Skolem Theorem, simply formulated, states that if we let Γ be a set of formulae in first-order logic such that there is an infinite α with α a model of Γ, then there are models β of arbitrary infinite cardinalities such that β is a model of Γ. This implies then that a theory with an infinite model has infinitely many non-isomorphic models.

[11] Poincaré's term originally.

CHAPTER FOUR

VARIATIONS ON THE NON-STATEMENT VIEW OF SCIENCE[1]

1. INTRODUCTION

In the Introduction I have briefly mentioned the basic differences between the statement and the non-statement approaches. The advocates of the statement approach depict scientific theories as axiomatised deductively closed sets of sentences within some appropriate syntactic system, and discuss the "empirical interpretations" of these theories in terms of some set of correspondence rules or bridge principles. The defenders of the non-statement approach, on the other hand, do not view the formal formulation of scientific theories in some appropriate language as the most useful characterisation of "theories". Rather, they depict these "theories" in terms of sets of mathematical structures that are the models of the theory in question.

In what follows I shall discuss a few of the main non-statement programmes offered by current philosophy of science. This is necessary since these programmes acknowledge the role of mathematical models in science's processes, which is of paramount importance to my model-theoretic view of science. Any kind of comparison is always limited and I do not claim aspects of the different accounts depicted, discussed, and compared here to be "exactly" the same. Rather I view these to be, at most "comparable" in a loose sense of the word, and offer these discussions more for clarification purposes than for making any naive claims of reducing any of these views to any other.

2. PATRICK SUPPES'S SET-THEORETIC APPROACH TO SCIENCE

Suppes offers one of the first viable alternatives to the "received (statement) view" of scientific theories and so brings about a radical turn in philosophy of science. Other than the structuralists who stress the use of formal semantics and meta science to appeal to the structural aspects of theories, Suppes finds the axioms of set-theory sufficient, and claims mathematics, rather than meta-mathematics to be the language of science.[2]

The class of structures (systems) that Suppes considers is described by giving one "generic" structure, with parameters, which can be specified to deliver all the systems in the class. Suppes (in Morgenbesser, 1967, p.60) acknowledges this when he points out that one of the simplest ways in which to provide an extrinsic characterisation of a theory is to define the intended class of models of the theory; and then asking if the theory can be axiomatised, merely comes down to asking if a set of axioms can be stated such that the models of these axioms are precisely the models in the defined class. He (ibid., pp.61, 62) remarks however that

> ... the problem of intrinsic axiomatisation of a scientific theory is more complicated and considerably more subtle .... Fortunately, it is precisely by explicit consideration of the class of models of the theory that the problem can be put into proper perspective and formulated in a fashion that makes possible consideration of its exact solution.

I agree, and, in model-theoretic terms, moreover, such consideration of the class of models of a given theory shows the continuous character of science (see Chapters 2 and 5).

Suppes addresses the philosophically problematic relations between empirical systems and theories (i.e. my "second set" of interpretational relations) in terms of a hierarchy of models that focuses on the complex nature of the experimental process.[3] He (Suppes, 1954, p.243) already points out very early on in his work that progress in foundational studies of philosophy of science requires distinction between theory and experiment, since the reconstruction of the experimental stage of science is rather more problematic in comparison to the theoretical stage which may be axiomatised "quite easily" with the help of set-theoretic predicates. He (ibid., p.246) wants to provide philosophy of science with "... a kind of algebra of experimentally realisable operations and relations" and emphasises that discussion of the empirical interpretations of the primitive notions for certain defined notions of some empirical theory imply interpretations of quantitative notions, which necessitates some systematic theory of measurement.

He is not interested in the classic notion of absolute objective truth, nor is he interested in the kind of framework offered by instrumentalists, rather he wants to speak more pragmatically about truth in terms of modern statistical decision theory. Thus, one of the most important issues in Suppes's philosophy of science is the emphasis he puts on the "experimental stage" of science.[4]

> ... the point of a theory of measurement is to lay bare the structure of a collection of empirical relations which may be used to measure the characteristics of empirical phenomena corresponding to the concept. Why a collection of relations? From an abstract standpoint a set of empirical data consists of a collection of relations between specified objects. For example, data on the relative weights of a set of physical objects are easily represented by an ordering relation on the set; additional data, and a fortiori an additional relation, are needed to yield a satisfactory quantitative measurement of the masses of objects" (Scott & Suppes, 1958, p.113).

Thus, as far as the co-ordinating principles or bridge principles of the statement approach are concerned, Suppes stresses (in Morgenbesser, 1967, p.62) that the practice of testing scientific theories is a much more complicated issue than is implied by the usual comment about these issues.[5] (See Chapter 3.) I agree with this, but I do not choose to turn to statistical methodology to examine these relations that Suppes (in Morgenbesser (1967), and also in Suppes (1969), Suppes (1989), and Suppes (1993)) insists on. For the purposes of a realist philosophy of science, it is sufficient — and a more philosophically challenging prospect, I might add — to analyse the various semantic model-theoretic relations involved.

I am, of course, not denying the use of statistical methodology to clarify and determine as precisely as possible the chances of a theory's models having connections to some systems in reality. I merely prefer a model-theoretic analysis of theories, combined with a(n adapted) Van Fraassenian notion of empirical adequacy and non-monotonic preferential analyses of empirical over-determination instead of the mathematical tools of statistics being employed to answer inherently *philosophical* questions about possible relations between science and reality. Also, I agree with Wójcicki (in Humphreys, 1994, p.130) that Suppes's set-theoretical position may reduce philosophy of science to no more than "selected problems of metamathematics". Wójcicki also remarks:

> Needless to say, as long as an empirical theory is not provided with any factual interpretation, it remains merely a certain formal system. But ... one may wonder whether the *differentia specifica* allowing us to tell an empirical theory from a piece of pure mathematics does not consist in the fact that the former has some intended empirical applications.

This was essentially Adams's (1959) argument, as Suppes also points out. He started the idea of an empirical theory consisting of two classes of structures, the one the class of all the theory's realisations and the other the class of all intended applications which is a class of empirical structures (physical systems) of which the theory is expected to be true. Adams also pointed out that not every intended application necessarily has to be a realisation of the theory. Sneed (see Section 4.3) modified these notions in the sense that he requires that no component of an intended application be T-theoretical, while Adams saw the intended applications of structures of the same set-theoretic type as the realisations of the theory itself, which is close to how model-theoretic realism also portrays it.

Suppes and Dana Scott in their article *Foundational aspects of theories of measurement* (1958) ground the foundational analysis of measurement in general model theory. Suppes (in Morgenbesser, 1967, p.58) points out that the essential characteristic of a theory of measurement is that it can study (in a precise way) the transformation or development of "qualitative observations" into the "quantitative assertions" characteristic of the more theoretical stages of the scientific process. He approaches this problem in terms of representation theorems, mainly because he views the models of the theory and the models of the data (see below) to be of different logical types. He (ibid., p.58) writes:

> Given an axiomatised theory of measurement of some empirical quantity such as mass, distance, or force, the mathematical task is to prove a representation theorem for *models* of the theory which establishes, roughly speaking, that any empirical model is isomorphic to some numerical model of the theory. The existence of this isomorphism between models justifies the application of numbers to things. ... What we can do is to show that the structure of a set of phenomena under certain empirical operations is the same as the structure of some set of numbers under arithmetical operations and relations.[6]

Although I would discuss the above in terms of empirical reducts and empirical models and although model-theoretic realism does not need representation theorems, this is essentially (as far as these views are comparable) my view of the "verification" of the models of scientific theories too. In the model theoretic context it is however not necessary to use a separate language — from the one talking about the content of a theory's interpretative models — to talk about the empirical models of theories[7] — although of course it can be done, and then Suppes's use of representation theorems will become applicable too.[8]

Suppes (1954, p.245) sets out the various stages of formulating a set-theoretic predicate for (or axiomatising) a particular branch of empirical science as follows:

- In the beginning some kind of statement of what other theories are assumed (e.g. in axiomatising rigid body mechanics, one would assume the standard branches of mathematics and particle mechanics) is needed.

- Then the "primitive" notions of the theory are listed, and their set-theoretic nature (in particle mechanics, notions like "set of particles", the "interval of elapsed time", the "position function", the "mass function", and so on) is indicated.

- The set-theoretic definition can then be completed by listing the axioms which have to be satisfied, because one will then be able to examine the deductive consequences of the definition.

Obviously one of the main tasks here is to rationally reconstruct within set theory the standard theorems of the branch being studied. One will also then be able to ask some of the questions of modern mathematics that have obvious implications for the structure of empirical theories, such as questions concerning the formulation of representation theorems which may be linked for instance to studies directed towards the problem of reduction between theories[9]. Then finally, one will be in a position to give an empirical interpretation of the axiomatised theory, which will have to take the complexity of the entire experimental enterprise into account.

Suppes thus articulates a more complex stratified view of the relations between models of theories and systems in reality, than I do in this text. However, I too stress the elaborate sophistication of the manoeuvres needed to find the possible links between particular systems in reality and certain models of a theory being examined at a given time, although he (Suppes, 1989, p.25) wants to show that the study of the relations between (empirical) theories and their data demands a study in terms of a hierarchy of models of different logical type[10]. He (Suppes, 1960, p.297) stresses the clarifying role of the set-theoretical notion of model in experimental design and the analysis of data:

> The maddeningly diverse and complex experience which constitutes an experiment is not the entity which is directly compared with a model of a theory. Drastic assumptions of all sorts are made in reducing the experimental experience ... to a simple entity ready for comparison with a model of the theory.

Suppes sees the empirical relation between an interpretation (of a given theory or class of systems) and a system in reality as a highly articulated, composite relation, with an articulation which depends upon the experimental or observational situation in question. Suppes (1989, p.25) explains that

VARIATIONS ON THE NON-STATEMENT VIEW OF SCIENCE   95

> Theoretical notions are used in the theory which have no direct observable analogue in the experimental data. In addition, it is common for models of a theory to contain continuous functions or infinite sequences although the confirming data are highly discrete and finitistic in character. ... Corresponding to possible realisations of the theory, I introduce possible realisations of the data. As should be apparent, from a logical standpoint possible realisations of data are defined in just the same way as possible realisations of the theory being tested by the experiment from which the data come. The precise definition of models of the data for any given experiment requires that there be a theory of the data in the sense of the experimental procedure, as well as in the ordinary sense of the empirical theory of the phenomena being studied.

He (ibid.) gives two reasons why a possible realisation of a theory cannot be a possible realisation of its data: no actual experiment can include an infinite number of discrete trials, and the parameter of the experiment is not directly observable and is not part of the recorded data. In other words, models of the experiment and models of the theory are of a different logical type. Note that although in a model-theoretic approach none of this is denied, this still does not necessarily imply that different languages are needed to talk about the content of the interpretative models and empirical models.

I thus agree with his hierarchical view, but for my (semantic realist) purposes I collapse this complex relationship to a much simpler relation, indicated by "empirical adequacy" (in my terms). This simple relation results in fitting the empirical data — however elaborately extracted from the physical system[11], and subsequently formulated conceptually, i.e. mathematically — into the relevant interpretative model of the theory in question (i.e. the relevant conceptualisation of the empirical data in question is shown to be isomorphic to a substructure (i.e. the empirical reduct) of the interpretative model in question). A simple example: Observations over time deliver 113 different spatial positions $(x_1, y_1), (x_2, y_2), ..., (x_{113}, y_{113})$ for the planet Neptune in the x-y-plane (of the planets in our solar system) of a coordinate system centred on the sun. These are the data. All 113 points *lie on* the (near-) ellipse with its uncountably many points which is the interpretative model of the orbit of Neptune (which in its turn is part of the interpretative model of the solar system, in which the — Newtonian or Einsteinian — theory of our solar system is true). What is suppressed or collapsed here is the process of distilling the data $(x_1, y_1), (x_2, y_2), ..., (x_{113}, y_{113})$, which is a process which involves theories, models, and practices relating to telescopes, the human eye and visual system, light, movement of the earth, clocks, and so on, and so on.

Suppes (ibid., pp.27-29) argues that the fundamental theory and descriptions of apparatus are two extremes of hierarchy — in between are the models of the theory, the models of the experiment, and the models of the data. He (in Morgenbesser, 1967, p.62) emphasises over and over again that to be able to "connect" experimental data to a relevant theory, the data have to be put through a "conceptual grinder", which refers to this conceptual hierarchy he sets out from the "raw" observations to the final "fundamental" scientific theory. The theory of the experiment is the definition of all the possible realisations of the theory that is the first "step down from the abstract level" (Suppes, 1989, p.28) of the fundamental theory. A possible realisation of the theory of the experiment is a model of the theory if the experimental conditions are satisfied. Models of the experiment represent experimental data in canonical form, but when is a possible realisation of the data a model of the data? Suppes (ibid., p.29) remarks again that an answer to this kind of question requires a "detailed statistical theory of goodness of fit", since models of the data should incorporate "all the information about the experiment which can be used in statistical tests of the adequacy of the theory" (ibid., p.31), which means that he (ibid., p.32) restricts the models of the data to those aspects of experiments which have a parametric analogue in the theory. In these terms, I would express Suppes's model of the scientific development of a theory (ibid., p.31) as follows:

Fundamental theory
Models of the fundamental theory
Theory of the experiment
Models of the experiment
Models of data
Experimental design
Ceteris paribus conditions

Suppes (ibid., p.32) characterises the *ceteris paribus* conditions at the bottom of the table as "every intuitive consideration of experimental design that involves no formal statistics", which I presume refers to the context in which the concrete as-yet-untranslated-into-data "first" observational activities are carried out. Note that in a model-theoretic account of theories the *ceteris paribus* conditions are taken to be active in the formulation of the theory and not at the level of dealing with data. He seems to think, as Cartwright in her turn also decides, that the closer to reality we get, the more of these clauses we need, while I claim that only by a suspension of these clauses can we move to the more specific levels of the scientific process. Thus although I do not deny the idealised character of our interpretative models or even of the images of real systems our empirical models present, I claim that these "idealisations" are not so much a result of *ceteris paribus* clauses as simply of the nature of scientific actions. (More about this later in this chapter and also in Chapter 5.)

A last point to note is that he (ibid.) makes it clear that in analysing the relations between theories and experiments, difficulties encountered at any of the levels from the level of models of the theory to his last identified level of *ceteris paribus* conditions reflect problems or weaknesses in the relevant experiment, and not in the relevant "fundamental" theory. I agree, although it seems obvious that errors may also occur in the formal construction of the models of the theory, and also, that problems encountered in creating these constructions may indeed point to problems in the structure of the fundamental theory[12].

Suppes (ibid., p.34) concludes:

> One of the besetting sins of philosophy of science is to overly simplify the structure of science. ... What I have attempted to argue is that a whole hierarchy of models stands between the model of the basic theory and the complete experimental experience. Moreover, for each level of the hierarchy, there is a theory in its own right. Theory at one level is given empirical meaning by making formal connections with theory at a lower level.

"Stretched out" my version of a model-theoretic view of the structure of theories can accommodate something like this kind of hierarchy. There is a principle of transitivity at work here — accommodated by a model-theoretic account of theories — that does the same philosophical work as Suppes's intricate comments on measurement theory and the role of representation theorems. Take Newton's laws of motion and his law of gravity applied to our solar system again. Such a model (i.e. a model of our solar system) will be described by an uncountable set of sentences which, i.a., describes every position at any time of every planet in question on an elliptical curve. I claim there exists a transitive connection between experimental data and some model of the theory, since the data offer "pieces" of the model by means of some "experimental theory" in the sense that the theory of the experiment "translates" observations into data (or models of data), which can be then possibly linked with (i.e. embedded into) some (reduct of some) model of the theory. If a scientist is looking at a planet through a telescope, the theory of the telescope translates those observations into data, and these data give the position of say Mercury at a given point in time. But this is exactly what is depicted by some empirical model of Newton's theory embedded into some reduct of the particular interpretative model (in this context), since the latter model of the solar system offers here the positions of *all* the planets at (all) specific times. The data thus do depict certain relations valid in models of the theory.

3. THE STRUCTURALIST PROGRAMME

Stegmüller ((1976), (1979)) places the structuralist programme in the "non-statement" tradition, because its main methodological principle is to view "theories" as *structures*, or sets of structures (in the standard set-theoretic sense), in the place of sets of statements. Sneed (in Balzer, Moulines, Sneed, 1987, p.86) writes

> ... our point of view is not that of a philosopher who is puzzled by questions like how it is possible to obtain knowledge about 'the phenomena'. Our attitude is much more descriptive, and if we look at what is 'given' for an empirical theory we take the point of view of that very theory — in contrast to the 'absolute' point of view of the philosopher.

This refers to the fact that the structuralists are more concerned with making apparent the logical form or structure of empirical theories than with their actual content, and also illustrates Stegmüller's instrumentalist tendencies. That is why their aim is to create "logical reconstructions" of empirical theories, rather than worry about the "usual philosophic issues to do with science and reality" (Sneed, 1983, p.350)[13]. Also, Sneed (1971) and Stegmüller (1976) claim this approach can best, among other things, handle Kuhn's condition for membership of a particular scientific community in terms of "holding" a certain theory, although a community's beliefs about the subject matter of that theory is not fixed over time.

Structuralists typically depict a scientific theory as a "... *conceptual structure* that can generate a variety of empirical claims about a loosely specified, but not completely unspecified, *range of applications*" (Sneed, 1976, p.120). This view is closely related to what I hold too, except for the fact that a scientific theory to me still is a linguistic expression — and an essential component of science — that may be interpreted by a set of (conceptual interpretative) structures (i.e. models in which the theory's sentences are true). Like Sneed, I also take these models (and not the theory itself) to be the "generators" of "a variety of empirical claims" about a certain range of applications of the theory.

The structuralist programme dates from the early sixties and is essentially a development of Patrick Suppes's view. The programme originally started (Sneed, 1983, p.350) as an attempt to describe more precisely the empirical claims of theories with considerable mathematical apparatus. Since the early seventies Wolfgang Stegmüller, with his colleague Joseph Sneed, assisted by Wolfgang Balzer, Ulises Moulines and others, started refining this approach. The structuralist approach is a meta-theoretical approach to scientific (empirical) theories that essentially focuses on the nature of scientific theories, interrelations between theories (especially reduction, equivalence, and approximation), and theory progress or evolution. During the last two decades in various disciplines formal reconstructions of empirical theories into structuralist form have been carried out by the supporters of the programme, for example in physics (for instance Sneed (1973)), psychology (Suppes (1969), (1989)), and neuropsychology (Suppes (1989)), and also in economics (for instance Balzer (1982), (1985)).

Before I briefly discuss the main tenets of their programme, a comment on their view of the role of language, given their "non statement" approach. In *An architectonic for science* (Balzer, Moulines, Sneed, 1987, p.17) it is claimed that

> ... we [Balzer, Moulines, and Sneed] believe ... that in the study of the structure and development of empirical science language has not a big role to play. This belief is mirrored in our concept(s) of an empirical theory which will not contain a language as an explicit part. It would ... be a serious misunderstanding to say that therefore we have dispensed with language altogether. ... What is left open in our account is the way in which sentences are formed out of basic symbols, variables, and other logical symbols. But such formation rules do not play any role in empirical science ... .

They (ibid.) go on to point out that a species of structures totally describes the "non-logical" vocabulary of a given theory (i.e. its individual constants, predicate constants, function constants and their arities).

I agree that the last point may be true, but I have a problem with the first remark claiming that the formation rules of sentences (of theories) do not form part of empirical science and that this is justification enough for the structuralists to leave it out of their analysis of empirical theories. Of course these rules (like the rules of mathematics) do not form part of empirical science itself. However, (again like mathematics) they provide the means to formulate and communicate the conceptual structures, and they do form part of a philosophical analysis (or rational reconstruction) of science, since part of such an analysis would be to examine the structure and meaning of scientific theories as linguistic expressions. This is where the real choice between the statement and the non-statement approach lies.

Retaining the notion of theories as linguistic expressions is the only way I can see in which to formulate the class of models of a theory in "one" formal expression. Even the structuralists have to employ language to describe the classes of structures they have in mind, albeit the mixture of natural language and mathematical symbolism usually employed by scientists and mathematicians, and of which a formal logical language is a (meta-) mathematically amenable stylisation.[14] Also, as mentioned before, the linguistic expression of a theory offers different ways of "controlling" the multiplicity of models of one theory in the sense that amending the axioms of a given theory can logically strengthen the theory such that it "shrinks" the class of possible models (via boundary or initial conditions), or the theory can "shift" to a different class of models, such as happened when Einstein added the "cosmological constant" to his equations (theory!).

Joseph Sneed, in giving an exposition of what he sees as the subject matter and nature of philosophy of science (Sneed, 1976, p.121), explains that the structuralist claim is that

> ... everything interesting a scientist of science might want to say about the products of science can be said within [a] conceptual [structural] framework. More precisely, it can be said within an ontology of scientific theories ... with sufficiently ingenious relations among these entities.

He remarks (ibid., p.116) that philosophy of science in general is about setting out a "clear, coherent conceptual framework" in which the various sciences can formulate their empirical claims. However, he (ibid.) continues that the "science of science" he is thinking about should be a social science, since its main subject matter are the communities within which these empirical claims are formulated. The coming-into-being, development, and going-out-of-existence of these communities as well as of their products (empirical scientific theories) should, it seems then, all form part of this subject matter.[15] A model-theoretic realism takes the latter into consideration too, while saying what it wants to say within an ontology of science ("theories") rather than that of reality, illustrating both that it is not a metaphysical scientific realism of the kind Putnam refers to, and that it does not have to become (part of) a "social science" to accommodate certain contingent (social) factors in the development of science.

The basic elements of theory identification as set out by the structuralist programme may be summarised as follows (note that structuralists claim that all these components can be precisely explained in purely structural — i.e. set-theoretical — terms). "Theories" are taken to consist of (classes of) models in the Tarskian sense of formal semantics, i.e. a model of a theory T is a possible realisation in which all the sentences (or at least a set of given axioms) of that theory are satisfied.[16] The identity of a "theory" is first and foremost given by a class of models, which we may call M (following Stegmüller and his colleagues).[17] The models are determined by a given set of axioms (the "tautologies" of the theory), but the structuralists claim these axioms to be secondary to the determination of the identity of a theory, since any set of axioms may be chosen just as long as it is satisfied by the same set of models, M.

However, although it is true that in the structuralist view the specific set of axioms in question does not really play a primary role in the identification of the theory, distinguishing between two different kinds of axiom does. In the literature, these two types of axiom are usually referred to (Moulines in Schurz & Dorn, 1991, p.317) as *framework conditions*[18], which are mainly the accepted body of theories, or background knowledge, or paradigm within which scientists work, and "*proper*" *axioms* which are taken to be laws (containing theoretical terms). In the example of Newtonian mechanics, Newton's Second Law is a "proper" axiom, while the (implicit) condition of the differentiability of the position function would be a framework condition. The framework conditions define the basic notions about the structure of each of the fundamental notions of the theory, i.e. the "base set" (Balzer, Moulines, Sneed, 1987, p.5) of the theory; while the "proper" axioms state the law-like relations between these basic notions, i.e. "what conditions have to be satisfied for a possible candidate to really be a structure of this kind?". The latter are the "fundamental laws" (ibid., p.19) that "connect" all the terms of a theory in one "big formula".

Structures determined only by framework conditions are called *potential models*, the class of which may be denoted by M_P, or $M_P(T)$, while structures fulfilling both the framework conditions and proper axioms of a given theory, are called *actual models*, the class of which may be denoted by the familiar M, or $M(T)$. $M(T)$ is a subset of $M_P(T)$ — so that $M \subseteq M_P$. Methodologically speaking, clear distinction between these two classes of models is necessary in any logical reconstruction of a theory[19]. Thus the set-theoretic predicates determining the set $M_P(T)$ are defined by statements about the set-theoretic properties of the base set of the theory, and typifications and characterisations of basic relations. The determination of $M(T)$, on the other hand, relies on the specification of the laws (axioms) identified in the theory as well[20].

The structuralists try to "fit" models to theories, since they do not acknowledge the theory as a linguistic entity to start off with. A defender of a model-theoretic account of science, in her turn, works towards the formulation of a theory as a linguistic expression, and then, in applying the theory, starts off with the theory as a linguistic entity and tries to construct models in which that particular theory will be true. Also part of the structuralist determination of the identity of a theory is a distinction between the theoretical and the non-theoretical terms of the same theory, the interrelationships ("constraints") between models of the same theory, as well as the inter-theoretical links between models of different theories (concerning different sets of potential models). (More about these a little later in this section.)

Advocates of the structuralist programme take $<M_P,M> = K$ (Moulines in Schurz & Dorn, 1991, p.319, and Balzer, Moulines, Sneed, 1987, pp.36ff.) to be the (conceptual) "theory-core" of a particular theory. The core K plus the class of intended applications, call it I, form the simplest set-theoretic structure that may serve as a logical reconstruction of an empirical theory. K and I are called theory elements. K is a purely formal mathematical structure and it says "something" or is "about" the class of intended applications. More complex theories are "built" of theory-elements that are linked or related in certain ways. In summary, the notion of a theory core is expanded to include the following elements (ibid., p.37):

- conceptual framework conditions,

- empirical laws, i.e. the "proper axioms",

- "constraints" describing connections or relations between different applications of a particular theory, and

- inter-theoretical links (such as the relations of reduction, approximation, and equivalence) describing links between a particular application of the theory in question and other different theories represented by different theory elements.

- Also part of the theory core is the class of partial potential models M_{pp} that consists of a subset of M_P, the class of potential models, that can be interpreted independently of the theory in question. Partial potential models are thus characterised in terms of a *theory-relevant* theoretic/nontheoretic distinction among the components of the class of potential models. More about partial potential models a little later in this section.

- An empirical claim — also part of the theory core — is associated with a particular theory element in terms of the part of the content of that theory element which forms the class of partial potential models that is "compatible" with the laws, constraints, and inter-theoretic links associated with the particular theory element in question. This claim is simply the claim that the particular intended application of the theory in question is in K, i.e. in $<M_P,M>$.

An obvious motivation (that both realists and anti-realists would agree on I should think) for empirical theory construction surely is the (successful) application, in one way or the other, of that (empirical) theory. That is why claiming that we know what an empirical theory looks like if we know its core, is not completely correct. We also need to have some information on the nature of its intended applications. Structurally speaking, then, if we take I as the set of intended applications of a given empirical theory identified by a specific given K, we have to know the nature of the elements of I, as well as the extension[21] of I. Note again that cores of theories and the applications of theories together — i.e. M_P, M, and I — are the "material" out of which empirical claims may be formulated.

Now, the elements of I are taken — by the structuralists — to be not "simply the 'real things', independent of any conceptualisation, to which the theory is supposed to apply" (Moulines in Schurz & Dorn, 1991, p.319)[22], but rather systems, which are nothing else but structures, that present us with ways of "... conceptually carving up reality in pieces and putting these pieces in certain relationships" (ibid., p.320). Thus, we can take a system, s, to be a structure of the form $<A_1, ..., A_m, R_1, ..., R_n>$. The important issue here is to determine the relationship between the class of potential models, M_P, and a particular intended application, s. Obviously, for a system s to be an intended application of a theory, it has to be an element of the set of potential models, M_P, in the core of that theory, and thus a necessary condition for the determination of a set I of intended applications of a theory given b☐$K = <M_P, M>$, becomes $I \subseteq M_P$[23]. Sneed (in Humphreys, 1994, p.196) points out that I should be seen as the "totality" of potential data the theory in question is supposed to account for.

All right, but exactly what does this subclass (I) look like? How big is it? To be able to understand the structuralist answer to this question, let us think a bit about the nature of the relations between I and the class M of actual models in K. We cannot simply assume that $I = M$, since it is entirely possible that any empirical theory might have applications that are unwanted, for whatever reason. The implicit result of multiple models of empirical theories entailed by the depiction of theories in terms of classes of models[24] lies in the problem of identifying "empirically uninteresting" models (or potential models) so that they may be discarded (as soon as possible). As this is a problem defenders of the statement approach also face, it is especially interesting to see how the structuralists face up to it. Balzer, Moulines, and Sneed (1987, p.23) offer an explication of the application of a theory to a certain (intended) range of phenomena in terms of M_p and M. They first point out that this issue is related to a distinction between conceptualising a certain range of phenomena and making an empirical assertion about them.

Moulines (in Schurz & Dorn, 1991, pp.321, 322) sets out three possible ways to describe the nature of the relations between M and I:

- (i) $I \subseteq M$
- (ii) $I \subsetneq M$, but $I \cap M \neq \Phi$
- (iii) $I \cap M = \Phi$.

Case (i) is the ideal case, where the relevant theory is a complete success, capturing all the intended applications. Case (ii) presents us with a partially successful theory, and, of course, the bigger the intersection between I and M, the more successful the theory will be. Case (iii) presents us with a theory that has no intended applications among its actual models, which, of course is a completely meaningless theory.[25]

Moulines (ibid.) points out that in terms of the methodological evaluation of empirical theories it is evident that the intersection between I and M needs to be defined or specified as precisely as possible, although the structuralists stress that there is *no purely semantic answer* to the question concerning this intersection.[26] In a realist model-theoretic context such as mine, this answer is given by the isomorphic relation determining empirical adequacy and also by the non-classical preferential analysis of empirical model choice (see Chapter 2).

Now, according to Moulines and his colleagues, any kind of approach to this issue has to be preceded by what they term "pragmatic-diachronic considerations" (ibid., p.321), because of the fact that for every given theory core, K, there has to exist a *scientific community* that will *use* the theory identified by the core in "real life". Because I is dependent on the scientific community within which the theory under consideration has been constructed or will be applied, the structuralists refer to the class of intended applications as a "genidentical" (ibid., p.322) entity.[27] The relationship between the scientific community and the pair K, the core (identity) of the theory in question, is philosophically important, because intended applications are taken — by the structuralists but not by me — to be part of and internal to the theory concept itself, and not somehow external to the theory. Balzer, Moulines, and Sneed (1987, p.38) claim that without this class it will be impossible to know the empirical content of the theory. They (ibid.) put the structuralist case as follows:

> We consider [the class of intended applications] to be a part of the identity of a theory because without it we would have no way to know whether we are dealing with an *empirical* theory at all. Take the case of an advanced scientific theory, where quite a few abstract terms expressed in mathematical language appear, and let us ask whether by just considering the theory's formalism we would be able to tell which part of the world the theory describes, or for what purposes the theory is useful. ... we cannot tell. For, even by assuming that the formalism is adequate for describing some part of the world, we should be able in general to go over to quite different phenomena described by the same formal

means. This is indicated by a well-known theorem of logic, namely, that structures isomorphic to models of a theory are again models of that same theory. [Reminding of Putnam's warning in terms of his "model-theoretic paradox" (Chapter 3).] ... Thus, in order to know what a theory is about, we have to include an informal description of its intended application, as a part of the identity of the theory in question.[28]

The relationship between the set of intended applications and a given scientific community can vary in the sense that the same community may use completely incompatible cores. On the other hand, different communities may use the same core class and then the relationship between intended applications and cores will be completely different. This is possible because we can define (from the *intended* nature of these applications) the domain of intended applications of a certain core K_i, associated with a specific scientific community, as a particular subclass of the class of potential models, M_p, for which the scientific community in question wishes (intends) to show (either by observation, experiments, or calculation) that it is also a subclass of the class of actual models, M, of the theory in question.

> Strictly speaking [of course], I may not be characterised as a class in the precise sense of set theory since it is not determined by purely extensional means. ... The domain I remains always within M_p, but its precise limits *within M_p* change as the skills, knowledge and interest of [the scientific community] as a whole change. (Moulines in Schurz & Dorn, 1991, p.324)

This reminds of what I termed the problem of "over-determination" in Chapter 2, and has also been pointed out earlier by Suppes, and as I have pointed out above, not even the original formulators of some scientific theory can know in advance what will happen to their theory — i.e. in which models it will be interpreted. In my terms, this is a matter to be determined via the empirical models of the theory about the construction of which nothing can be said beforehand. And the only thing that can be said about a scientific theory's "identity" in the above structuralist sense is the remarks I made about the role of the "intended models" in the developmental stage of theories. This is no problem though, since given the nature of scientific knowledge, the nature of these applications simply has to be open-ended in this way.

Balzer, Moulines and Sneed (1987, p.23) offer the following explication of the application of a scientific theory to a certain (intended)[29] range of phenomena in terms of the set of potential models and the set of actual models of the theory. They write:

> When confronted with some given 'data' or 'phenomena' we might want to use a theory T to 'understand' them, to 'explain' them, to 'predict' them — in short we might want to apply T to these data. To do this, the first thing we try is to conceptualise the domain I of data in terms of T, i.e. [we] ... use the concepts appearing in potential models of T to refer to I. We create a potential model of T for I. This is the more 'conceptual' aspect of the application of a theory. The next step is to make an assertion about I in terms of T — an assertion with empirically testable consequences. We then assert that I satisfies the fundamental laws of T, which, of course, only make sense if I has already been conceptualised in terms of T. In other words, we make the empirical assertion that the potential model considered is also an actual model of T. This empirical assertion can be either true or false. If it turns out to be true, we can say that we applied T to I successfully.

More specifically, Sneed (1976) and his colleagues (Balzer, Moulines & Sneed (1987)) describe the set of (intended) applications of an empirical theory in terms of some set of partial potential models (i.e. theory-*independent* subsets of the set of *potential* models), and an empirical claim associated with the core of the theory in question. Such an *empirical claim* states that the set of partial potential models that satisfies the conditions set by the laws, constraints, and inter-theoretic links of the theory in question, is indeed in K, the theory's core. If one recalls that the class of partial potential models represents subsets of the class of potential models, the above explication in terms of intended applications remains the same in general.

This finer distinction of the class of potential models focuses on those theoretical terms that are specified by other theories and not by the theory in question. This implies that the part of a theory that may have relations to reality cannot be determined only by the theory itself. Sneed (in Balzer, Moulines, Sneed, 1987, p.86) remarks that the class of partial potential models represents what is "given" for a particular theory in terms of surrounding theories. This is the case because the set of partial potential models is what remains if all (T-)theoretical components are removed from M_p. This fact obviously fits well with Stegmüller's instrumentalism. Niiniluoto (1984) suggests that, if the intended applications of a theory are taken as *potential* models (which still contain theoretical entities), rather than suggesting the complicated structuralist relationship between partial potential models and intended applications, a realist interpretation of structuralism remains possible. But, model-theoretically it is the special (semantic) nature of the relationships between theories and their models, models and their reducts, and reducts and empirical models that imply realism.

Why is this distinction between so-called "T-theoretical" and non-theoretical terms necessary though? Let us first look again at the way in which the defenders of the structuralist programme see theories connected to their empirical claims. The structuralists' answer to these questions become very complex and extremely technical since they formulate these answers in terms of constraints, inter-theoretic relations, and partial potential models. Most simply put, as noted above, cores of theories and the applications of theories together form the "bricks" out of which empirical claims may be formulated.

Theories consist of these basic "theory-elements"[30] which have the ability to construct various "theory-nets" (which are basically more complex theories than the original one in question) from the relevant theory's original elements. A theory-net consists of a "specialisation" of the basic theory element. This specialisation corresponds to an empirical claim about the range of intended applications offered by the theory element in question. Thus the net as a whole corresponds to a non-basic empirical claim about the whole range of intended applications the theory has to offer. The problematic part of this analysis of theories lies in the overlaps between applications in various theory elements. These theory elements are linked by the inter-relationships between models of the same theory (referred to as "constraints" in the structuralist programme). And, it is these problems that may be solved — or so the structuralists seem to think — by distinguishing between theoretical and non-theoretical terms.

More specifically, Sneed's answer to the questions surrounding the question of theoreticity is close to the criterion Kuipers (2001) uses to denote epistemological stratification, i.e. a criterion referring to the theory in which the concept under discussion appears. Kuipers (ibid.) offers a more simple formulation than Sneed's for a general distinction between two kinds of "non-logical or mathematical" term in relation to a statement S, but here I shall explain the more general formulation of so-called T-theoretical-ness as Stegmüller (1979, p.116) following Sneed sets it out.

Stegmüller (ibid.) summaries this criterion as follows: "... a quantity f is theoretical relative to a theory T iff the values of f are calculated in a T-dependent manner". Stegmüller stresses the pragmatic implications of Sneed's criterion when he remarks (ibid., pp.117,118) that it may be viewed as a

> ... partial explication of the phrase 'meaning as use': The extensional meaning of the term 'force' in classical particle mechanics consists of the values of the force function. And the assignment of these values depends on how laws including forces are used. [This does] ... not ... [imply] an identification of the meaning of the term 'force' with just the use of this term, but rather with the use of the (general and special) laws containing the term.

The structuralist emphasis on the use of laws determining the latter's empirical extensions confirms their instrumentalist sympathies. Such emphasis, if needed, may however be viewed in realist terms by fitting it into the default framework for choices for empirical models sketched in Chapter 2.

The consequence of the application of this "T"-criterion to the structure M is a "decomposition" (ibid., p.118) of M, as follows: The class M_P is, as explained above, the class of possible models

> ... of the full conceptual apparatus, for which it remains open whether the 'actual axioms' are valid for them. (In most cases M will only form a small subset of M_P.) The set of all entities that are retained if one removes all theoretical components from M_P is called the set M_{pp} of ... partial potential models. In classical particle mechanics, eg., they are systems of moving particles, described only in spatio-temporal respect, but not endowed with forces or mass.

Thus the further class of partial potential models M_{PP} is obtained by taking the elements of M_P and for each of them forming what we could call — following Kuipers (2001) — an "observational reduct". Recall that a "reduct" in model-theoretic terms is created by leaving out in the language and its interpretations some of the relations and functions originally contained in these entities. In the structuralist case it is relations, functions, and constants which correspond to T-theoretical terms that are left out to define such a reduct.

The T-criterion is expanded by the structuralist notion of constraints which are simply laws in general philosophy of science terms. Stegmüller (1979, p.118) explains:

> It appears that, eg., certain features of special force laws as well as the ... unity of mass-ratios and extensivity of mass with respect to particle concatenation are most naturally treated as relations among different models of classical particle mechanics. Formally the difference between laws and constraints is expressed by taking laws as subsets of M_P and constraints as subsets of $\wp(M)$.

If the set of constraints is called C, K now becomes K = <M_P, M_{PP}, M, C>.

Recall that M represents the so-called "fundamental" laws which holds for every application of a theory. Sneed proposes that "special" laws should be identified with subtheories of the main theory, retaining the same structure as the original theory.

Stegmüller (ibid., p.119) comments:

> This reconstruction happens to correspond also to physical usage, according to which one speaks alternatively of the law of gravitation or the theory of gravitation. For an ordered pair <K', I'> being a special law has the following necessary conditions: $M' \subseteq M$, $C' \subseteq C$, $I' \subseteq I$. ... what originally was called 'theory' loses its special status as soon as laws are constructed according to the same model. [Now] ... we speak only of theory elements instead of theories. The original theory is now distinguished as the basic element. The indicated process of forming special laws which allows iteration is called specialisation. With the help of specialisation one can construct a whole net of theory elements with the basic element on top. The specialisation relation furnishes a partial order of the theory elements.

Stegmüller (ibid., pp.122,123) points out that this "enrichment" has the result of transforming theory elements into quadruples of the form T = <K, I, SC, h>, where SC is a particular scientific community and h is an historical interval such that SC intends to apply K to *I* during h.

In this sense the definition of the core of a theory is expounded (Sneed, 1976, p.123) as follows: $K = <M_P, M_{PP}, M, C>$, where

- M_p is the set of potential models of the theory, as in the above, but with the understanding that these models are models of the *entire* conceptual content of the theory, including theoretical components;

- M_{PP}, the set of partial potential models, is the set of all models obtained by excluding the theoretical components from the conceptual body of the theory in question[31];

- *M* is still the set of actual models, since it depicts the set of possible models of the "full conceptual apparatus" of the theory that satisfy certain laws formulated in terms of theoretical components;

- *C* is the set of constraints on M_P, and captures the notion that different applications of the same theory-element are interdependent[32] in the sense that values of a function in one application of the theory may not be used without taking account of the values of that function in other applications — thus, constraints single out certain admissible combinations of potential models[33]; and

- inter-theoretical links are depicted as follows: T-nontheoretical terms may be also or only determined by means of other theories which do not presuppose T at all. Balzer, Moulines, and Sneed (1987, p.58) write:

> The information [from theories different from T that we want to transfer to T] consists of data which are obtained in the course of some determination of a term which is *non-theoretical* in T. Clearly such transfer contributes to the interpretation of $M_{PP}(T)$. It is part of the determination of the meaning of the terms occurring in T's partial potential models, and therefore it is an essential component of T itself. ... [This leads to the introduction of] ... inter-theoretical links which represent the transfer of data from theories T' to theory T.[34]

Sneed (1976, p.124)[35] claims that a theory-element core is used to make empirical claims in the sense that a subset of M_{PP}, call it $A(K)$ (ibid.), is selected such that theoretical elements can be added to each of its members in such a way that it yields a subset of the set of actual models M (this means that each member of the subset of M_{pp} will satisfy the theoretical laws of the theory).[36] The empirical claim that the particular theory element is thus making, is that descriptions of phenomena that actually occur is indeed a part of the theory core. In other words, if we have a theory-element $E = <K, I>$, where K is the elaborated theory core above, and I remains the set of intended applications, then the claim that E is making is that I is an element of the subset $A(K)$ of M_{pp}.

That means that the theory-element core K narrows down the set M_{pp} to the subset $A(K)$, thus restricting the possible models of the theory (containing only non-theoretical components) such that the result is I. This "narrowing down" is done via constraints and inter-theoretical links[37]. Balzer, Moulines, and Sneed (1987, p.87) point out that the assumption that the intended applications of T have the structure of its partial potential models is the "most economical and most natural" assumption to make. Thus we should assume that $I(T) \subseteq M_{pp}(T)$. There will, of course, still be unwanted applications, even if we take M_{pp} as the set of all possible applications of the theory, but these gentlemen (ibid., p.87) claim it is enough that "[w]e *can* say something precise [after all], namely that an intended application is a partial potential model, but we cannot be precise about *every* feature of intended applications".[38]

In other words the empirical claim (or hypothesis) corresponding to the structures above is formulated as follows:

> First, we form for some theory element the intersection $\wp(M) \cap C$. Further we use the restriction function r: $M_p \to M_{pp}$, which through cancellation produces an 'empirical system', that is a partial potential model out of every potential model. The analogous function, which operates on classes of M_p's is called [r*]. The application operation A, applied to a theory element core K, is to be defined by $A(K) := [r*](\wp(m) \cap C)$. A subset of M_{pp} is in this set if and only if theoretical functions can be added to its elements in such a way that one obtains a subset of M and that in addition this array of theoretical functions satisfy the constraints C. To every theory element then corresponds the empirical assertion made by the scientific community SC during h: $I \subseteq A(K)$. (Stegmüller, 1979, pp.122,123).

The whole empirical assertion should be viewed as a conjunction of the assertions $I_i \subseteq A(K_i)$ for the corresponding elements of the abstract theory net.

In Kuipers's (2001) terms this comes down to the fact that within the class of partial potential models (which is a subclass of the class of potential models) lies the class ΠM of the observational reducts of the structures in the class of actual models, M. Also in the class M_{pp} lies I, the class of intended applications. The empirical claim associated with a certain theory then, is that I is a subset of ΠM, or at least that the intersection between I and ΠM is non-empty. The question to be asked now is whether this implies that the structuralist theoretic/observational distinction might be as naive as the positivist one, in the sense that they do not relativise their reduct to particular applications of M. Surely more than one reduct both of the class of potential models and of the class of actual models exist, depending on both the real system at issue and the nature of the classes M_p and M, since non-isomorphic models may have isomorphic empirical substructures — so the structuralist reduct projections may be many-to-one (or even many-to-many) — without any harm done either to (moderate) realist ideals or to theory-observation disentanglements (if model-theoretic relations between theories and their models, models and their reducts, and reducts and empirical models are assumed).

A final point on the nature of the set of intended applications. Gerard Schurz (1995, pp.279-280) in an article entitled "Theories and their applications - a case of non-monotonic reasoning" points out that the structuralist set-theoretical claim that $I \subseteq A(K)$ implies that "every application which belongs to I will satisfy the theoretical claim of theory core $[K]$". He also (ibid., p.280) remarks that in the structuralist reconstruction the "description of the intended applications is not a part of the theory reconstruction". Well, perhaps not in the sense that they do not view the refinement of the intersection between I and the set of partial potential models as a semantic issue, but rather in pragmatic terms, but obviously they do claim (Balzer, Moulines, Sneed,1987, p.38) that without the class I it will be impossible to know the empirical content of the theory. They (ibid.) write:

> We consider [the class of intended applications] to be a part of the identity of a theory because without it we would have no way to know whether we are dealing with an *empirical* theory at all.

In model-theoretic terms describing the possible empirical models that may be embedded into some reducts of some models of the theory might perhaps by some be viewed as part of theory formulation in terms of the "original" intended models(s) (see Section 2.3) of the theory, but in the context of theory application, such descriptions are not part of theory reconstruction in the sense that in this context these models have to be *found* to be "embed-able" as it were into some reduct of some model of the theory.

Schurz (ibid.) continues that obviously "... structuralists realise the importance of the parameter I because only via I the theory core becomes related to the real world. But they just assume I to be given in some pragmatic way". Schurz (ibid.) claims that in science actually the intended applications of a theory form an essential part of the theory. Although Stegmüller (1986, pp.28ff.) comes to the conclusion (already mentioned above) that I must be an open set and that no sufficient conditions for limiting I can be given, structuralists view I as an "extensionally fixed set (though no one knows how it becomes fixed), and the relation between a theory and its application is understood as a deterministic one" Schurz (1995, p.281). Schurz (ibid.) points out that if some conflict exists between the empirical claim of a theory and data, then the set I, the entire theory element, and even the theory net have to be changed, which leaves us facing difficult questions regarding theory succession. He (ibid., pp.285-292) views an application of non-monotonic logic as a way out, as I do, although both our motivations for this and the nature of our applications differ somewhat (see Chapter 2).

4. THE SEMANTIC APPROACHES OF BETH, VAN FRAASSEN AND SUPPE

Bas van Fraassen developed a semantic approach to philosophy of science by building on the work of Evert Beth[39], in which physical systems are depicted in terms of their possible states. This position was further developed by Frederick Suppe[40]. The foundational motivation behind this approach is that any scientific (physical) theory is taken — by scientists themselves — to have many alternative linguistic formulations. (Think of the Lagrangian and Hamiltonian formulations of classical particle mechanics.) Theories thus cannot be identified with their linguistic formulations. Suppe (1973, p.130) claims that it is rather the case that

> ... scientific theories are extra-linguistic entities which are referred to and described by their various linguistic formulations ... [thus] theories are to be constructed as abstract *structures* which serve as models for the sets of interpreted sentences [that] constitute their linguistic formulations (i.e. that they are *meta-mathematical models* of their linguistic formulations), where the same structure (theory) may be the model for a number of different, and possibly inequivalent, sets of sentences or linguistic formulations of the theory.

Here, then, is a more radical approach than either that of Suppes or the structuralist programme in the sense that the theory is *identified*, as it were, with the notion of model. A theory *is* a model to the defenders of this approach, they do not merely talk about discarding the linguistic features of theories in logical reconstructions, they claim a theory to be "extra-linguistic".

Beth developed what is referred to as a "state-spaces" view (related to the older phase and configuration space view of mechanics and that of Von Neumann (1955) for quantum mechanics) in three articles (1948/49), (1949), and (1961). Van Fraassen (1970, pp. 337,338) writes:

> Beth's approach does not require or presuppose the complete formalisation of the theory under analysis. ... His approach takes into account the essential role of models in science. In Beth's account, the mathematics is not part of the physical theory, but is used to construct the theoretical framework. The theoretical reasoning of the physicist is viewed as ordinary mathematical reasoning concerning the framework. ... Finally, Beth's approach makes possible the use of formal semantic concepts and methods.

Van Fraassen claims (ibid., p.338) Beth's approach to be much closer to the actual foundational work done in physics than any variations of the statement approach. Van Fraassen (ibid., p.327), following Beth, believes that the meaning structure of a certain part of natural language becomes suitable for a technical role in some scientific language if it has a representation in terms of a model, in the sense of a mathematical structure. Then the scientific language can be formally reconstructed as an artificial language whose semantics is determined with reference to this mathematical structure or model. Such a language Van Fraassen (1967) calls a "partial or semi-interpreted language".[41]

Notice that the "meaning structure" of a part of natural language that may be represented by some mathematical model may here then be described in some appropriate formal language. In a model-theoretic approach the semantic content of the linguistic expression of the theory in some appropriate formal language is determined by the initial interpretative model, but also, the linguistic theoretical expression is then interpreted by other interpretative models during the application stages of the theory[42], which implies nothing more than that the "meaning structure" of the linguistic expression is then represented yet again by other (or the same) mathematical structures. However, this is not what Van Fraassen and Beth really claim. To them the model is the mechanism that may determine the semantics of the formal language in which the theory may be formulated, as are my initial interpretative models. However the theory itself remains a non-linguistic entity, since nowhere do they mention the possible interpretation of the theory in terms of other mathematical structures in the model-theoretic sense.[43]

Now, in Beth's approach, the notions of a "physical system" of a theory and the various "states" in which this system can be at given times, are foundational. These notions are however easier to understand if some of Suppe's notions are introduced first. The specific class of phenomena that the (linguistic) formulation of a theory is meant to characterise, is called the *intended scope of the theory*. Theories do not characterise these phenomena in their complexity, though. Suppe (1973, p.131) gives as illustration the fact that classical particle mechanics characterises mechanical phenomena as if they depend only on the abstracted position and momentum parameters, while actually various other "unselected" parameters usually also influence the phenomena. Thus a theory's characterisation describes what the relevant phenomena *would have been like had* the abstracted parameters — those the theory's formulation focuses on for whatever reason — been the only parameters influencing them. This is close to what happens in interpretative models' empirical reducts of a given scientific theory. More about this a little later in this section.

In this sense, theories may be said to characterise *physical systems*, because they are *about* the behaviour of certain abstract systems in the sense that this behaviour is dependent only on the parameters selected by the theory. Physical systems are relational systems whose domains consist of states, and relations and laws ranging over these states.[44] Van Fraassen (1970, p.330) states that the function of a law in Beth's approach is to describe the behaviour of the physical system with which the theory is occupied at the time in terms of its possible states, its normal evolution through time, and its behaviour in interaction with other factors.[45] These laws thus indicate which states are physically possible for the various physical systems; and they also determine which combinations of states are so-called theory induced physical systems (notion explained below) and which are not.

> Thus the relations of the theory determine all and only those sequences which are the behaviours of physical systems in the class of theory induced physical systems. (Suppe, 1973, p.133.)

The selected parameters abstracted from the phenomena can wholly describe the behaviour of physical systems, and so they are called the *defining parameters of the physical system*. The values of these parameters are physical quantities which may be determinate or statistical (ibid., p.131). A set of simultaneous values for the parameters of a physical system is a *possible state* of the system. Note that any physical system is at any time in exactly one of its possible states, although that state may change over time. The behaviour of a physical system is given in terms of these state changes. In this way, the behaviour of a system is the system's history and each physical system has a unique sequence of states (in the deterministic case) or a set of possible sequences of states with associated probabilities (in the statistical case) that it assumes over time. Each physical system is characterised fully by a specification of the possible states it can assume and the sequences of states it assumes over time.

The class of *causally possible physical systems* for a theory is the class of physical systems which correspond in the following way to causally possible phenomena *P*, within the theory's intended scope: any *P* in the theory's intended scope corresponds to a (causally) possible system *P'* such that *P'* is what any causally possible phenomenon *P* would have been were the idealised conditions imposed by the theory met and the phenomenon *P* influenced only by the selected parameters. Obviously then, one of the tasks of a theory is to describe the class of causally possible physical systems for the associated theory. This is done by the theory describing a class of physical systems known as the *theory induced class of physical systems*, such that this class is identical to the class of causally possible physical systems.

A theory then is *empirically true* if the theory induced class of physical systems and the class of causally possible physical systems for the theory are indeed identical. Testing of theories involves determining whether this identity in fact exists between these two types of classes of systems and is usually done in statistical terms.

Suppe (ibid., p.136) elaborates on Beth's discussion of the usage of propositions[46] and uses this elaboration to discuss the semantic relations between theory-formulation languages, theories, physical systems, and phenomena. He then defines a *formulation* of a theory as a set or class or "collection" of propositions which are true of the theory. Suppe (ibid., p.161) explains that he uses the term 'proposition' in the sense of a linguistic entity that may be propounded in the sense that the component sentences of a proposition do not fully determine the proposition since these sentences may be used to express various different propositions.

> Roughly speaking, when a proposition is asserted with reference to some subject matter it becomes a statement. As such propositions are interpreted declarative sentences which may be propounded with reference to one or more situation .... (Ibid.)

A theory formulation usually consists of a few specified propositions with all deductive consequences of the specified propositions under some "logic". These propositions are in a language called the "theory-formulation language", and usually forms a subset of the propositions of that language. The following basic features of a theory formulation may be identified (ibid., pp.137,138):

- A set of elementary propositions in the theory-formulation language specifies that a certain physical parameter p has as value a certain physical quantity q at time t, such that an elementary proposition ϕ is true of state s in the theory's domain if at time t, s has q as the value of the parameter p.

- For each elementary proposition ϕ there is a maximal subset $h(\phi)$ of the theory's domain such that ϕ is true of all the states in that subset.

- The function h from elementary propositions to subsets of the theory is called the *satisfaction function* for the set of elementary propositions.

- Elementary propositions may be compounded together in accordance to the *logic of the theory* — the logic is such that every compound proposition is true of at least one state which is — according to the associated theory — physically possible; thus obviously the logic of a theory is theory-dependent and different theories may have different logics. (Suppe's (ibid., p.137) illustration: classical particle mechanics impose a Boolean algebra mod-2 and quantum theory imposes a non-distributive lattice.) In Beth and Van Fraassen's terms (Van Fraassen, 1970, p.335) the logic of the theory is essentially a syntactic description of the set of valid sentences and the semantic entailment relation in that language[47].

- A *language of description* is determined by the set of elementary propositions, the theory, the satisfaction function h, and the logic of the theory; this language is obviously a sublanguage of the theory-formulation language; this language can describe any physically possible state in a physical system.

- It might be possible that the logic of the theory enables one to deduce logical consequences of propositions in the language of description. However, usually the language of description has to be incorporated into a more complex language with an amended logic — namely none other than the theory-formulation language — which can express the laws of the theory and deduce predictions.

- The truth conditions for the theory-formulation language are specified in terms of the relations (laws) of the theory and the truth conditions for the language of description.

- A formulation of a theory is a set of propositions deductively closed under the logic of the theory-formulation language such that every proposition in the set is true of the theory.

- Finally, the theory-formulation language may be a natural or an artificial language, but typically is a language such as "scientific English".

For analysing semantical relations holding between propositions in the expanded theory-formulation language and phenomena, Suppe (1973, pp.140ff.) offers an operationalist account of factual truth that I shall not discuss here. Van Fraassen (1970, p.328) writes, as far as the semantic content of the theory formulation is concerned, that the "set of *states* of some physical system are represented by elements of a certain mathematical space, called the *state-space*."[48] Apart from the state-space, these theories use a certain set of parameters — referred to in the above — to characterise the particular physical system. This yields the theory's set of elementary statements about the system in question. These are initial and boundary conditions which are part of the relations determining empirical adequacy in my terms. These statements are such (ibid.) that each elementary statement U formulates a proposition to the effect that a certain physical magnitude m has a certain value r at a certain time t.[49] The truth of such a proposition U always depends on (or is relative to) the particular state of the system at that time — in some states m will have the value r and in some states it will not have that value.[50]

So, from the above, it seems that a "satisfaction function" determines whether the system's actual state is represented by an element of the mathematical structure

VARIATIONS ON THE NON-STATEMENT VIEW OF SCIENCE   113

consisting of all the relevant state-spaces of the theory or not:

> The mapping h [of the theory] (the *satisfaction function*) ... connects the state-spaces with the elementary statements, and hence, the mathematical model provided by the theory with empirical measurement results.[51] ... The exact relation between ... [elementary statement U] and the outcome of an actual experiment is the subject of an auxiliary theory of measurement, of which the notion of 'correspondence rule' gives only the shallowest characterisation. (Van Fraassen, 1970, p.329.)

A description of a set of state-spaces plus the satisfaction function are thus offered in the place of the statement approach's axioms or postulates concerning the "primitive" symbols of the scientific language.[52] Van Fraassen's semi-interpreted language thus in these terms consists of the elementary statements connected to a certain physical system, the specific state-spaces in question, and the satisfaction function in question.

Thus, to summarise, in Suppe's terms the theory's domain is the class of states of the system possible according to the theory formulation, corresponding to the structuralist class M of actual models. The truth of our proposition φ always depends on (or is relative to) the particular state of the system at that time — in some states p will have the value q and in some states it will not have that value.[53] For each elementary proposition φ, there is a maximal subset h(φ) of the theory's domain such that φ is true of all the states of the subset. The function h from elementary propositions to subsets of the theory ('s domain) is called the satisfaction function for the set of elementary propositions. The satisfaction function thus embeds elementary empirical facts into the mathematical model of the theory formulation in question.

If a physical system is in the class of theory induced systems, then the domain of the physical system will be a subset of the domain of the theory and the sequence of states of that system will be one determined by the theory's relations (laws). These physical systems are meant to be *replicas* of the actual systems in reality, and so by describing the physical systems the theory "... indirectly gives a counterfactual characterisation of the actual phenomena" (Suppe, 1973, p.131). Also, as implied above, it may happen that theories give an *idealisation* of some physical system — (ibid.).

Such idealised physical systems are still abstract replicas of phenomena, but with the additional feature that certain idealised conditions (such as being isolated systems of dimensionless point masses) are imposed on these systems *which actual phenomena can never actually meet*. Thus Suppe (ibid., p.139) writes:

> Only some of the propositions which are true of the theory will be true of a particular physical system in the class of theory-induced systems, but every proposition true of a physical system in that class will be true of the theory.

If a theory is empirically true, the semantic relations holding between propositions in the theory-formulation language and the class of causally possible physical systems for the theory will be the same as those holding for the theory-formulation language and the class of theory-induced physical systems. Moreover, every proposition in the theory-formulation language which is true of a causally possible physical system will be true of the theory.

A few remarks on this semantic approach to scientific theories in terms of a model-theoretic account of these matters. In model-theoretic terms the intended scope of a theory is indeed the "class of phenomena" the (formulation of) the theory is meant to characterise. Within the latter kind of account we usually speak of *real systems* of phenomena or of classes of phenomena. Also, in the latter approach, a theory can indeed do no more (as Nancy Cartwright so delights in pointing out) than characterise the real system it means to describe as it would have been had the abstracted parameters of the theory been the only ones influencing the system in question. (I shall discuss this "idealised" or "open-ended" feature of scientific theories more deeply later on in this chapter and in Chapter 5.)

Each possible state of a physical system — in Suppe's terms — might be viewed as an interpretative model in model-theoretic terms, seeing that a possible state of a system is given in terms of a simultaneous set of values for the parameters of the theory in question. The theory-induced physical systems may perhaps be viewed in terms of the empirical reducts of a model-theoretic account. The reason for this is that the class of causally possible systems (i.e. empirical models in my terms) turns out to be systems in which the idealised conditions set by the theory have been realised, influenced only by the selected parameters of the theory, and that a theory is said to be empirically true in Suppe's terms if this class of systems is identical to the class of theory-induced physical systems. The relation of empirical adequacy between interpretative and empirical models of some theory in model-theoretic terms is then very close to this relation of identity between causally possible physical systems and theory-induced physical systems. This becomes plausible if we take into account that the "satisfaction function" of Suppe's semantic framework determines whether the actual state — i.e. the causally possible state — of a physical system is represented by the mathematical structure representing the theory-induced physical systems.

Returning briefly to Van Fraassen's constructive empiricism (Van Fraassen, 1980), he denies the need for dwelling on questions concerning reference between theoretical entities and aspects of reality. According to this view, relations of empirical adequacy between phenomena and "empirical substructures" of models of theories that refer to "observational" terms of theories, are sufficient for the needs of philosophy of science. To see where Van Fraassen and I part ways, let us briefly examine the distinction that Newton made between "apparent motion" and "absolute or real or true motion" as Van Fraassen (1976, pp.624ff., 1980, pp.44ff.) sets it out.

In Ptolemy's terms the earth is stationary. In Copernicus's terms the sun is stationary. In Newton's terms neither the sun nor the earth is stationary. Planetary motion in Newtonian celestial mechanics is observed relative to the earth's motion. The notion of apparent motion is introduced such that the apparent motion of particular bodies accounts for the differences of their "true" motions (Van Fraassen, 1976, p.624). Ptolemy need not have made the distinction between true and apparent motion, since to him "true" motion was exactly what was observed. In Copernican terms we can only observe the planets' motion relative to the earth, which is not stationary and thus the apparent motion of the planets are the difference between the earth's true motion and the true motion of the planets. Newton generalised apparent motion — i.e. motion relative to the earth — to the motion of one body relative to another. Any observed motion thus became a relative motion and an apparent motion is motion relative to an observer (Van Fraassen, 1980, p.45).

Newton separated the reality he "postulated" from the phenomena he "saved" (ibid., p.44) by referring to the "absolute magnitudes" of his axioms and to their experimental determination as "sensible measures" (ibid.). Apparent motions "form" (ibid.) relational structures defined by measuring relative distances, time intervals, and angles of separation. These relational structures Van Fraassen (ibid.) calls "appearances". In a mathematical model of Newton's theory bodies are located in Absolute Space in which they have "true" motion (ibid.). Van Fraassen (1976, p.624) writes:

> But within these models we can define structures that are meant to be exact reflections of ... appearances and are, as Newton says, identifiable as differences between true motions. These structures, defined in terms of the relative relations between absolute locations and absolute times ... I shall call motions ... .

The notion of "appearances" thus refers to the actual observed motions — the "phenomena" — while the notion of "motions" refers to the terms of some mathematical model of Newtonian celestial mechanics.

Van Fraassen (1980, p.45) continues "[w]hen Newton claims empirical adequacy for his theory, he is claiming that his theory has some model such that *all actual appearances are identifiable with (isomorphic to) motions* in the model". Thus, in Van Fraassen's and my terms, empirical adequacy — in Newton's terms — would mean that all empirical models (Van Fraassen's substructures) of Newton's theory will be isomorphically embedded into the particular (interpretative) model defined above. However, as both Van Fraassen and I remark, all that is really necessary or possible for determining the empirical adequacy of a given theory, is that it has at least one (interpretative) model with an embedded empirical model (adequately empirically related to some real system). I use this isomorphic link between empirical reducts of interpretative models and empirical models though as a kind consolidation of theoretical/observational distinctions, while Van Fraassen does not really seem interested to do this.

In model-theoretic terms the models of the theory should indeed be adequate to the phenomena, but if the fact that the *theory* may be "adequate" to (true in) its (interpretative) models is taken into account as well, we have a model-theoretic realism that addresses the possible meaning and reference of "theoretical entities" without relapsing into the metaphysics typical of the usual scientific realist approaches. Remember that the deductive structure of theories is mirrored semantically in the (interpretative) models of the theory and thus cannot be represented fully by looking only at empirical reducts (or even at Van Fraassen's substructures) of these models.[54] Surely in terms of the above, it is the appearances that allow us to make sense of the motions (via the relevant models) which in their turn allow us to apply Newton's *theory* in a certain way? Van Fraassen does not care about the answer to this question, while I do very much.

Van Fraassen (1980, pp.63, 64) also claims that physical theories describe "much more than what is observable" (ibid.). However it is still the empirical adequacy of the theory in question that really matters. He (ibid., p.64) claims that the notion of empirical adequacy does not "collapse" into a mere notion of (metaphysical) truth, since

> ... it relates the theory to the *actual* phenomena (and not to anything which *would* happen if the world *were* different, assertions ... which ... have, to my [Van Fraassen's] mind, no basis in fact but reflect only the background theories with which they operate). (Ibid.)

So empirical adequacy offers us a way in which we can delimit all the talk about the various possible models of some theory such that we can show one of these models to be actually about some real system. Van Fraassen of course would only state that it can be shown that some model is about the phenomena in some real system and not that this fact somehow links the theory to the relevant real system as well. Since in model-theoretic terms the models of a theory both interpret the theory and conceptualise the aspects of the real system in question, here it is argued that the property of empirical adequacy concretises not only (at least one of) the models of some theory, but also the theory itself. Model-theoretic empirical adequacy of a theory thus also does not collapse into a simple (metaphysical) notion of "truth" but rather in a more subtle way, it collapses into the notion of "articulated reference respecting the data".[55]

5. RONALD GIERE'S NATURALISTIC APPROACH TO SCIENCE

Giere (1985, p.75) agrees with Van Fraassen and the model-theoretic point that the logical positivists' (statement view) pre-occupation with the linguistic structure of scientific theories obscures the important role models in which those theories are true have to play in the scientific process. He, however, does not waste much time in pursuing any of the semantic categories of reference and meaning in the way Van Fraassen does (via his notion of elementary statements yielding semi-interpreted languages). He (ibid., p.77) states clearly that he

> ... will simply ignore such issues. ... the theory of science need not wait on the development of adequate general theories of meaning and reference to proceed. We need not know in detail *how* general terms such as *mass* come to be associated with terms in an abstract mathematical structure. We know *that* it can be done because it *is* done.

This is close to what I have been implying when I stressed that, because of the complex and changeable nature of these issues, questions concerning experimental design, measurement theories, and criteria determining the "fit" of some model to a system in reality, are finally left to science itself to answer. I do however think philosophy of science has *something* to say about these issues, at least as far as showing their place in the structure of science as a whole (i.e. in the relations between the possible empirical and interpretative models of some theory), and their implications for the structure of scientific theories in particular (i.e. their possible reference to real systems).[56]

Giere presents himself as a supporter of the non-statement view of theories, preferring Beth's state-space approach to Suppes's set-theoretic one. He states (Giere, 1994, p.277) that he interprets the model-theoretic approach to imply that

... theories include two sorts of linguistic entities. Some are predicates, which may have an elaborate internal structure, as, for example, the predicates 'pendulum' or 'two-body Newtonian gravitational system'. Others are statements of the form 'X is P' in which X refers to a real world system and P to one of the predicates, as in the statement, 'The earth-moon system is a two-body Newtonian gravitational system'.

These remarks might be viewed in terms of a model-theoretic approach's relations in its interpretative models — as the "predicates" Giere refers to — and such an approach's empirical reducts — Giere's "statements" concerning the "real world". In general, the structure of a theory consists, according to Giere (ibid.), of a "family" of models or predicates, where the linguistic structure corresponding to a predicate is a definition instead of an axiomatic system (as in the traditional statement approach).

Giere addresses these issues in terms of "theoretical models" (the models (or set of models) created by defining a certain real system), and "theoretical hypotheses" which are statements picking out similarities between theoretical models and some system in reality.[57] He (Giere, 1983, p.271) makes it clear from the start that theoretical models (as definitions of systems in reality) have no empirical content. They may, however, be used to make claims about reality via the theoretical hypotheses that identify elements of some theoretical model with elements of real systems and then claim that the real system exhibits the structure of the model in question. Giere (1984:11) acknowledges the idealised nature of models[58] by stating (Giere, 1985, p.79) that a theoretical model is not "a faithful replica in all detail" of the object modelled. He (Giere, 1984, p.12) goes on to explain that theoretical models also come in various degrees of specificity, but points out that no such thing as *a* maximally specific model exists. This kind of model is always relative in the sense that it is a model of a designated type; thus a model is always an idealisation of reality.

He (Giere, 1983, p.272) stresses though that a theory is not simply a general model. He (ibid.) blames scientists for thinking that theories have empirical content and accuses them of using the term "theory" to refer to a "more or less" generalised theoretical hypothesis asserting that "one or more specified kinds of system fit a given type of model". He (Giere, 1985, p.78) accuses the logical positivists of conflating two separate functions of a theory, namely to offer general interpretations of theoretical terms such as "mass"; and to provide the means of identifying particular instances of these terms. The two "functions" of theories that Giere refers to might be respectively viewed in model-theoretic terms as the "work" relations between theories and their interpretative models and between these models' empirical reducts and empirical models do. In other words the positivist "conflation" of these two functions are also "separated" in the model-theoretic context.

Giere, it seems however, wants actually to study "how we as human beings use abstract models in describing particular objects in the real world" — which reminds somewhat of Wartofski's approach[59]. He does this not by means of that favourite realist notion of approximate truth, nor in terms of Van Fraassen's (1980, pp.45ff.) notion of approximation in the sense of one model of a class of models fitting the real system, but rather by means of a particular notion of "similarity". Giere (1985, p.80) claims that theoretical hypotheses assert that "[t]he designated real system is *similar* to the proposed model in specified *respects* and to specified *degrees*".[60] He adds that the precision associated with any hypothesis is always less than or at most equal to the precision of the measurement techniques employed at the time.[61] Giere (1983, p.269) thus finds the rationality of science in the testing of "highly specified theoretical models against empirical data". He (ibid., p.270) writes:

> I agree with contemporary students of probability, induction, and the foundations of statistics that the individual hypothesis is a useful unit of analysis. On the other hand, I reject completely the idea that one can reduce the rationality of the scientific process to the rationality of individual agents. The rationality of science is to be found not so much in the heads of scientists as in the objective features of its methods and institutions [Yes!].

He also (like Suppes, Van Fraassen, and Suppe) reflects on the complex nature and role of data in the testing of theoretical hypotheses:

> ... in order to determine whether a proposed model fits the world one needs some information about the part of the world in question. But not all information is relevant. We will use the term data ... to refer to the special information that may be relevant to deciding whether the model in question does fit. (Giere, 1991, p.29).

Giere (ibid., pp.37-40) offers a detailed programme for the evaluation of theoretical hypotheses, but I will not go into the finer points of that here. He (1983, p.272) remarks that theoretical hypotheses can also vary from having a very simple form to being a very complex type of claim, and that (Giere, 1984, p.13) they reflect the level of specificity of the corresponding theoretical model. The simplest form of a theoretical hypothesis is a claim that a particular identifiable real system fits a given model. In model-theoretic terms this would correspond to the claim that an empirical model "obtained" "from" the relevant real system sits isomorphically embedded into a given interpretative model, making the latter empirically adequate for the real system (relative to the procedures delivering the empirical model). However, claiming for instance that our solar system is a Newtonian particle system (with a suitable set of initial conditions) involves the whole mechanical theory of the Newtonian system.

He (ibid.) notes that theoretical hypotheses can thus be more or less general in the sense of including more or fewer real systems of various kinds.

> Consider, for example, Newton's theory of celestial mechanics, Mendel's theory of inheritance, or the plate tectonic theory of the earth. On this account ... the difference between a 'hypothesis' and a 'theory' may be largely honorific. (Ibid.)

Well, in model-theoretic terms this is not the case. The difference in scope between these kinds of theoretical hypotheses rather refers to the kinds of structures it relates to each other. The "truth" relations between an interpretative model and its theory will necessarily be more general in scope than those between an empirical model of a theory and one of its interpretative models. It is also worth noting that theories then rather than theoretical hypotheses contain all the different possible histories of some real system that could result from different, but physically possible, initial conditions.

6. NANCY CARTWRIGHT'S "SIMULACRUM" ACCOUNT OF SCIENCE

An interesting change of direction is offered by Nancy Cartwright's approach to the theories of physics. She is not a non-statement defender, but neither does she really fit into the statement framework. She does retain a (syntactic) notion of a theory, in the sense that she often refers to sets of field equations as theories, but it is not always clear what her views are on the notion of theories as deductively closed sets of sentences. Her continued claims concerning the "falsity" of the fundamental laws contained in scientific theories seem to indicate that, like the advocates of the non-statement approach, she views the role that theories (as linguistic entities) play in the processes of science at least suspect. Thus, like mine, her account of science has statement and non-statement characteristics and also she addresses the issue of realism in various ways throughout her account. That is why the whole of the next (rather long) section is devoted to discuss — against the background of my model-theoretic account of science — the main points of her approach.

My model-theoretic analysis of science is a realist depiction of the nature of scientific theories and their development meant to make use of semantic properties of formal languages to address relations between scientific theories and (systems in) reality. Nancy Cartwright is one of the most influential philosophers currently writing on the role of models in the process of science. Although some of her work comes very close to the model-theoretic interpretation of science that I am offering — for instance she also views "good' models as having "real" empirical substructures — there are also noteworthy differences in our approaches (to be discussed below).

Cartwright's main claim is that the "fundamental laws" which are part of the content of scientific theories have very little or nothing to say about actual states of affairs in reality. She argues for this with the aid of two arguments:

- an instrumentalist, anti-fundamentalist strategy arguing against a "theory-driven" interpretation of the function of models in (philosophy of) science (discussed in this section), and

- a metaphysical argument offering a hierarchy of causalities, dealing at the highest level with the capacities of real things — representing a "patchwork of laws" — based on a notion of reality as not necessarily being ordered and structured, even possibly being "disunified" (briefly commented on in Chapter 5).

She (Cartwright, 1983, p.77) states that a fundamental law can be regarded true only "when it states the facts". Given my realist motivated analysis of science I want to defend the value of scientific theories as linguistic entities making claims (even if they are of a general nature) about real systems in qualified contexts (models), and so, by implication, I want to defend the value of fundamental laws as well. Briefly, Cartwright's argument, as set out very clearly by Spurrett (1999), is simply that either laws must state the facts or be false, and, since in situations where more than one law combine neither of the laws satisfies Cartwright's facticity requirement, in such cases (both) laws are false.

She claims in *How the laws of physics lie* (1983) that considering the truth of the (fundamental) laws of physics will force anyone to admit that almost all of these laws are strictly false, i.e. "lie", because they are valid only under certain circumstances or given certain conditions *that do not strictly hold in reality*. However, it is interesting to note that the implication ["Conditions" → "Law"] is (logically) true even if the "conditions" (the antecedent of the implication) are not satisfied. Therefore "inapplicable" would be more appropriate than "lying", which seems to imply "false" in Cartwright's context. In a way, this is the basis of my attack on Cartwright: if science is analysed within a model-theoretic realist framework, both the inapplicability of certain laws in certain contexts and their ideal nature can be articulated, explained, and formalised without endangering either the value of fundamental laws in science or realist motivations.

She quotes (ibid., p.9) Boltzmann's equation and the general equation of continuity used by Maxwell in his explanation of the motion in a radiometer as examples of fundamental laws and describes these laws as "general, abstract equations; ... not about any particular happenings in any particular circumstances."[62] I do not think that this description of fundamental laws is debatable. What *is* debatable, however, is whether this necessarily leads to the conclusion that fundamental laws have no links with aspects of the real world.

Phenomenological laws are complex descriptions of actual situations in very specific terms —

> what can be confirmed through tests and comparisons with observations are phenomenological laws — comparatively detailed descriptions of concrete situations, which because of their richness in detail, do not have great generality (sometimes called 'low-level' generalisations). (Ibid., p.129)

Cartwright (ibid.) claims that it is phenomenological laws that fulfil the "traditional role" of laws in the sense that they describe empirical regularities — which fundamental laws — because they are too general and much too simple — cannot do, since they cannot account for the actually observed variety in the behaviour of objects in reality. Fundamental laws do not have anything to say about "regularities" (constant conjunctions of events in Humean terms), because describing regular behaviour requires more and more complicated descriptions of the situation. The descriptive phenomenological laws thus have less and less generality and they can never be stated without exceptions, while fundamental laws "by contrast, are simple, general, and without exception" (ibid., p.157).

It seems that Cartwright claims fundamental laws to be explanatory of the content of phenomenological laws, and phenomenological laws to be descriptive of aspects of reality. Explanation and description are thus done at different levels of the scientific process. This is a very important point, and is also accommodated in my model-theoretic account of science, but it is not a point that necessarily scores any marks for the falsity of fundamental laws. On the contrary, it is rather a *supportive* point in a model-theoretic realist account of science.

It is claims like the following about fundamental laws that do not seem to be entirely acceptable from a model-theoretic perspective —

... fundamental laws ... do not hold for the most part, or even approximately for the most part, and conversely, those laws which are more or less true much of the time are not fundamental. (Cartwright, 1989, p.174)

The unease that such claims cause is not necessarily the result of what she says about the nature of these laws[63], but rather that they seem to imply that she is nostalgic for an absolute notion of truth in the sense of being worried about so-called "fundamentalism"'s impact on realism. She stresses that fundamental laws can — possibly and at most — explain the content of phenomenological laws by organising or classifying them, and that fundamental laws therefore do not describe the behaviour of real objects in the world. However, as Rueger and Sharp (1996, p.95) point out, fundamental laws are in this context still useful to her even though they are not "true descriptions" of real objects or their behaviour, precisely because they "organise and classify our knowledge in an elegant and efficient manner" (Cartwright, 1983, p.100). She creates the impression in *How the laws of physics lie* (ibid.) though, that she might view the fact that fundamental laws only serve to organise and summarise real phenomena as a particular weakness of these kinds of law, because she puts so much emphasis on the fact that "the cost of explanatory power is descriptive adequacy" (ibid., p.3), which seems to imply that the final cost of explanatory power is the loss of the truth of fundamental laws.

In my version of the scientific process, however, that is not a problem, and, I might add, neither should it be in hers, because we both accept and acknowledge from the outset that truth is a very local and limited notion, albeit in a more complex way than is ordinarily thought. In other words, the fact that she (ibid.:5) denies that "explanation is a guide to truth", surely is only problematic if one thinks of truth as a universal notion. So Cartwright is arguing against the notion that fundamental laws give true descriptions of real phenomena. And thus, she is also arguing against my notion of scientific progress, because although we both acknowledge the use of the notion of models to mediate between the concrete and the abstract, she still thinks that accepting some kind of realism with regard to fundamental laws means accepting an absolute notion of truth. Why else does she say that fundamental laws *lie*?

If she takes seriously the possibility of contextualising the "truth" of these kinds of laws with the help of abstract models, why then does she still argue for the falsity of fundamental laws as if it is not possible for her to be satisfied with a "localised" version of truth? She does, in a sense, seem to imply something like this in *Nature's capacities and their measurement* (1989), as well as specifically stressing pretty clearly in her article entitled *Fables and models* (1986), the fact that questions of truth are not necessarily questions of universality.[64] Also in *The dappled world* (1999) she states that it is fundamentalism and not realism that is the enemy. So, perhaps we should see her linking universality (also in terms of truth) with fundamentalism, and the kind of truth she talks about in *Fables and models* (1986) and implies in *Nature's capacities* (1989) with possibilities of realism.

She argues against assumptions of the absolute truth of fundamental laws by stressing the concrete character of phenomenological laws. However, she cannot acknowledge the semantical links between theories and models that I claim exist, because she apparently thinks that would somehow imply that she believes theories to be absolutely true, and as they are not, she would rather discard them completely as part of the meaningful (and descriptive) side of the scientific process, and simply acknowledge (à la Duhem) their organising role, than try to find (like I am) some kind of reason for them to be part of the chain of factors or concepts that in the end make science mean (and explain) something to realists. (More on these issues a little later on in this section.)

Returning to Cartwright's interpretation of "phenomenological", she does not use it in terms of its usual interpretation as referring to the "observable" (Cartwright (1983); Cartwright (1986)), but rather points to the fact that this kind of law describes *actual* behaviour of *real* objects. She believes that, regardless of whether an object is observable or not, if we can manipulate it (intervene in its behaviour à la Hacking (see his *Intervening and representing* (1983)), we can formulate (true) low-level generalisations which accurately describe the (causal) relations into which it enters. Phenomenological laws describe *particular* events while whatever fundamental laws have to say is always about *various* situations in reality in one sweep. So, then — because fundamental laws can supposedly do no more than explain the content of phenomenological laws (in accordance with the covering law model of explanation, about which Cartwright has quite a lot to say), and good explanations are supposed to be simple (abstract) and general — it seems that fundamental laws can indeed never *directly* be about any *particular* aspect of reality.

But how then does Cartwright conceive of relating fundamental with phenomenological laws and either (or both) of these sets with real objects? It seems that the "content" of fundamental laws is filled in by various abstract models.[65] In Cartwright's scheme of things[66], these models mediate between theories and fundamental laws on the one hand, and phenomenological laws and reality on the other. According to the model-theoretic interpretation of the process of science that I am offering, models mediate between theories (linguistic systems) and systems in reality. Phenomenological laws, in my terms, would simply be part of the content (or properties) of the models interpreting scientific theories, and they would be expressible as sentences true in the reducts (given these structures' "empirical formulation") of the model(s) under consideration, as well as possibly true in some empirical models, and so, "true" of some real system.[67]

Schematically, Cartwright's account will look something like this:

SETS OF FUNDAMENTAL LAWS
THEORIES
↓
MODELS
Ceteris paribus conditions active
↓
Ceteris paribus conditions active
PHENOMENOLOGICAL LAWS
↓
REALITY

My scheme of things would rather be:

AXIOMS
FUNDAMENTAL LAWS
THEORIES
Ceteris paribus conditions active
↓
INTERPRETATIVE MODELS
↓
EMPIRICAL REDUCTS
↑
EMPIRICAL MODELS
↑
SYSTEMS IN REALITY

Figure 3

Cartwright explains in *How the laws ...* (1983, p.4), that the "route from theory to reality is from theory to model, and then from model to phenomenological law", and goes on to claim that "phenomenological laws are indeed true of objects in reality — or might be; but the fundamental laws are true only of objects in the model". In other words, she does not see the same kind of referential relation between models and systems in reality that I see. The reason, I think, is that she worries too much about the ideal character of the models and the role of the *ceteris paribus* clauses needed to interpret phenomenological laws. Also, it is difficult to see how a *very specific* link with reality can be given by a law, even if it is a phenomenological one. For example, if Newton's laws are fundamental, Kepler's are phenomenological (and deducible from Newton's), *but* the direct observations (done in both cases) are specific activities (looking in a particular precise direction) carried out at a specific time (specific to the second). Statements describing these kinds of activity surely are not *laws*, but can rather be expressed in terms of some empirical model which would be isomorphically embedded into a subset of the model of the theory under consideration and which interprets experimental data and empirical activities leading to the formulation of these data.

Cartwright is not an out-and-out anti-realist, as her interpretation of phenomenological laws clearly shows[68], it simply seems to be the case that she cannot see how to escape some of the seemingly anti-realist implications of the abstract nature of fundamental laws. This state of affairs has its origin in her interpretation of the "explanatory" role fundamental laws play in the practice of science (physics). It seems as if Cartwright is implying — in the sense of the validity of fundamental laws being dependent on abstract, idealised situations — that fundamental laws must hold *regardless* of the individual arrangements of things possible in each separate situation in reality that they (these laws) are "about"; while phenomenological laws potentially *describe* the *actual* situations to which they are applied.

This is all well and acceptable, but the problem is that Cartwright does not acknowledge that the sense in which fundamental laws may be said to hold "regardless of individual arrangements" is merely in terms of their abstract nature. This should *not* be linked to thinking that therefore they are universally true. Nor should it be thought that because they are too general to describe real systems, they are false. Questions of truth and reference can only be addressed in terms of theories' interpretative models.[69] In a model-theoretic realist account of the scientific process, as remarked above, I show that talk about the truth (or validity) of fundamental laws *per se* is unacceptable, and I argue that these issues can be meaningfully addressed only in terms of the infrastructure of the models interpreting these laws.

But, if then in Cartwright's terms, the main distinction between fundamental and phenomenological laws is taken to be the fact that fundamental laws hold by themselves — albeit only in certain "unreal" situations — while phenomenological laws can only hold on account of some (non-necessary) arrangement of circumstances, what does that imply for scientific explanation, prediction and the description of real objects? Phenomenological laws describe actual events, because although they are usually mathematically formulated in physics, no fundamental explanation of the mathematical formulae nor of the mechanisms underlying these formulae are assumed in these laws.[70] The problems related to scientific explanation in the context of the "leap" from fundamental and even phenomenological laws into more messy "worldly" situations are emphasised differently in a model-theoretic approach. Within such a model of science — as I have pointed out before — it is usually taken that scientific theories explain the content of their models, and through these models, some aspects of reality and the behaviour of certain phenomena may be described and predicted.

A theory does not always *necessarily explain* every detail of the system in reality it is focussing on. Newton's mechanics does not explain the phenomenon of gravity. It rather explains the influence of gravity on certain events and in that sense, it describes gravity rather than explains it. The old (deductive-nomological) symmetry between explanation and description should be "stretched" such that it covers all three strata of a model-theoretic model of science. If this is not done the fact that the descriptions of gravity in the above sense may enable someone applying Newton's mechanics to make certain predictions concerning the results of the exertion of the forces of gravity, without explaining gravity itself, cannot be grasped, and then it might seem that explanatory power indeed diminishes descriptive power, as Cartwright so often claims.

Thus, in model-theoretic terms scientific theories are said to *explain* in the basic sense of theories explaining the content of their models by establishing deductive links between the sentences expressing what is true in some model. Thus in a model-theoretic account of science a theory and its interpretative models "explain" in the strict logical sense that a predicted phenomenon can be logically deducted from the theory and the model(s) in question. Newton's three laws of motion and his law of gravity plus the model of our solar system — in terms of current scientific knowledge — explain why we see Mars tonight at eight o'clock in a particular position. In these terms, a preceding theory (e.g. Newton's laws of motion and gravitation) may describe models which (under certain conditions, within a certain interpretation, approximately) are also models of a later "higher order" theory (say the general theory of relativity), and then the latter may be said to explain the former.[71] The better explanatory power of later theories with respect to the content of their models is then the result of at least the higher level of accuracy of the theory. For instance, as Penrose (1997, p.57) points out, Einstein's general theory of relativity can be said to be accurate to about one part in 10^{14}, which is about ten million times as accurate as Newton's mechanics, which may roughly be taken to be accurate to about one part in 10^7. Improved accuracy is one embodiment of the continuity and progress in science with which some form of model-theoretic realism sits comfortably. (See also Chapter 2.)

The referential relations between model terms and objects and relations in some real system are (indeed, as Cartwright claims) more descriptive than explanatory. However, this need not result in anything as negative as Cartwright's claims of high explanatory power of fundamental laws diminishing their "truth making" power. If the whole model-theoretic realist interpretative chain — i.e. from terms of some theory, to terms in some interpretative model(s) of the theory, to terms in some empirical substructure of the interpretative model in question, to some real system — is taken into account, the fact that models seemingly describe and theories explain only the content of their interpretative models does not necessarily have any negative consequences.

The distinguishability and interconnectedness of the three stages roughly outlined by this "interpretative chain" — as set out in Chapter 2 — show however that description (a feature mainly of models) and explanation (a feature mainly of theories or fundamental laws) are inseparable, perhaps even just as much as explanation and prediction have traditionally been taken to be. Just as nothing can really be said about a theory's truth or reference without linking the theory to a specific interpretation of the relevant language given by some model of the theory, explaining something means at some level describing certain aspects of that thing. Definitions have to terminate at undefined terms, and the deduction of sentences of the theory has to start at unproven axioms.

126   CHAPTER 4

Cartwright has a valid point in emphasising the role of phenomenological laws against the overwhelming philosophical attention that fundamental laws get in so-called "fundamentalist" accounts of science, but model-theoretic realism differs from her account, because I introduce the role of models (and the laws — if any — active in them) from a different angle than she does. Because model-theoretic realism makes much of the role of fundamental laws in the scientific process, and also is (at least partly) categorisable in the semantic (non-statement) tradition, I find her account of fundamental laws limiting of (at least model-theoretic) realism and not only of fundamentalism. Unfortunately, she has few kind words to say to supporters of the non-statement approach to scientific theories as is obvious from this remark against defenders of the semantic account:

> On the semantic view, theories are just collections of models; this view offers then a modern Japanese-style automated version of the covering-law account that does away even with the midwife [of deduction]. (Cartwright, Shomar, & Suárez, 1995, p.139)

I agree that the non-statement elimination of the theory as a linguistic expression is misguided, and that is why in model-theoretic terms I stress the role of theory as much as I do the role of models. Theories (or fundamental laws) do indeed, in a certain sense, aim to "state the facts in a more general way so as to make claims about a variety of different circumstances" (Cartwright, 1983, p.103). But, I see them as a crucial link in the realist chain of scientific progress, and I stress that it is mainly thanks to their general nature in the above sense, that there are such links at all.

Laymon (1989, p.355) formulates one of the challenges Cartwright directs at supporters of the explanatory priority of fundamental laws as follows — they have to specify how actual scientific practice can be viewed as supportive of the truth of fundamental laws. Realists (in Laymon's terms) explain the practice of looking for increasingly more accurate and less idealised initial or boundary conditions in terms of the fact that idealisations are characteristically false and that they therefore have a distorting influence on derivations of predictions in such a way that, even if the fundamental laws are true, they will be able to produce only distorted or false predictions. In this sense, Laymon (ibid., p.359) claims, realists see the aim of science as the construction of more accurate models because they believe that our theories, if true, will produce more and more accurate predictions when applied to these more and more accurate models. He (ibid.) gives the example of Baily's connection of coefficients rendered superfluous and corrected by Stokes's development of a Newtonian theory of viscous fluids to illustrate his point. Cartwright (1983) gives a few examples in quantum physics to illustrate this realist tendency about approximation among "fundamentalists" — most notably Messiah's hydrogen atom (ibid., pp.137-138), and Louisell's treatment of the gas laser (ibid., pp.146-148).

Well, is giving idealised (perhaps "approximately true") descriptions (or explanations) of real systems not at least one of the things that science "really" is about? The trick is perhaps to distinguish between "distorting" idealisation and counterfactual idealisation. Think again of Suppe's distinction between counterfactual and ideal truth. Models provide us with counterfactual truths in the sense that they realise certain selected parameters of the theory in specific contexts where certain other influences are held stable. However the physical systems thus described are causally possible, while the kind of distortion to which Laymon and especially Cartwright seem to be referring, is Suppe's notion of "pure abstractions" that is not part of the empirical truth of models. (See Section 4.4 again for a discussion of Suppe's notion of empirical truth.)

These factors lead Cartwright however to conclude (ibid., p.107) that the empirical content of the phenomenological laws is *not* contained in the fundamental laws which supposedly explain them, in other words, she concludes that the fundamental laws simply organise scientific knowledge by explaining phenomenological laws but cannot really describe — or say anything else for that matter about — the real objects in the real world. Chalmers (1987, p.87) writes:

> Cartwright takes on an anti-realist stance with regard to fundamental laws, then, because the situations described by them are too simple and artificial to correspond to real world situations [no description of real objects], because adequate descriptions of the latter cannot in general be logically deduced from fundamental laws in conjunction with initial conditions [against the covering-law model], and because physicists frequently employ fundamental laws in diverse ways to offer different descriptions of the one real world situation [under-determination of theory by data].

She (Cartwright, 1983, pp.147-150) does however distinguish two senses of the notion of the "realistic" nature of a model:

- The first sense has to do with the relation between a model and reality (some aspect of reality, I would say). In this sense, a model is realistic if it gives an accurate description ("picture") of the aspect of reality ("situation") being modelled. In other words it will have to describe the structure and actual behaviour of the real system.

- The second sense has to do with the relation between the model and "the mathematics" (in her terms (ibid., p.150)), which is the relation between the model and the theory in my terms. According to her a fundamental theory determines criteria for what counts as explanations, and, in these terms — relative to those criteria — a model will be realistic if it explains the mathematical representation — i.e. if it realises the theory.[72]

She (ibid., pp.151ff.) offers her "simulacrum account of explanation" in the place of the covering law model of explanation. According to her (ibid., p.151) the covering law model requires the way in which phenomena are modelled to be realistic in both senses because it views a phenomenon to be explained if it has been derived from some fundamental law. Cartwright, however, primarily wants to show that — and how — fundamental laws logically summarise and classify (as mentioned before, in Duhem's tradition) groups of phenomenological (experimental) laws without *aiming* to explain them. She (ibid., p.152) writes:

> I have been arguing ... that the vast majority of successful treatments [of phenomena] in physics are not realistic. They are not realistic in the first sense of picturing phenomena in an accurate way; and even in the second sense, too much realism may be a stop to explanatory power, since the use of 'phenomenological' [still abstract] terms rather than more detailed 'causal' constructions may allow us more readily to deploy known solutions with understood characteristics and thereby to extend the scope of our theory [although this will not necessarily lead to a better understanding of the actual aspect of reality the fundamental laws are 'about'].

Cartwright's problem with fundamental laws is that they are laws about *distinct* (separate) aspects of objects in reality and their behaviour — or, in her most recent terms, about distinct causes and their separate effects — while, in the real world, these things actually occur only *in combination* with other aspects of these or even other objects. And, moreover, these combinations change quite often and occur very seldom according to some regular kind of pattern, because of the variety of factors involved. Cartwright's problem is that "[e]ven if these regularities did hold *ceteris paribus* — or, other things being *equal* — that would have no bearing on the far more common case where other things are *not* equal" (Cartwright, 1989, p.177). Again, this interpretation of the nature of fundamental laws is not really what is at issue here, the problem or challenge really is to find a kind of view that can accommodate these fundamental features of scientific theories and still offer a realist interpretation of the scientific process. Model-theoretic realism holds this promise without even having to specify whether one sees objects and activities in reality in terms of causes and their separate effects (as Cartwright seems to be doing nowadays) or not, since in model-theoretic realism the focus is on the semantics of science's language.

Now, if giving a fundamental theoretical account of a certain object means fitting it into the mathematical framework of the theory under discussion (as Chalmers (1987, p.84) remarks), and, if this is what fundamental laws ultimately do, as Cartwright claims —

> To explain a phenomenon is to find a model that fits it into the basic framework of the theory and that thus allows us to derive analogues for the messy and complicated phenomenological laws that are true of it. (Cartwright, 1983, p.152)

— the obvious question to me is *why* this should result in false fundamental laws? The answer lies in Cartwright's notion of models and their role in the scientific process.

She (ibid.) points out that models help us to "see" the relevant phenomenon through the mathematical framework of the theory, but stresses that different problems will have different emphases on different aspects of that framework. This, to me, implies that different models can — and should — only be evaluated according to the different aims guiding their construction (back, among other things, to the advantages of non-monotonic logic (in terms of a minimal model semantics), Chapter 2). And that is, in my view, why she calls her account of explanation a "simulacrum" account.[73]

However, as Rueger and Sharp (1996, p.95) claims, she seems to think that being useful in many different contexts,

... requires the theory to neglect the special differences between the contexts ...Therefore, the theory cannot be true of any of these *real* situations; it can give a correct description only of the behaviour of objects in highly idealised contexts or *models*. The model contains the distortions and idealisations that are necessary to make a theory bear on a real situation. Real objects and their behaviour are too varied, too complex, too messy to be treated faithfully by theories of great generality; that's why we need models to mediate between theory and phenomenon.

The important thing that both Rueger and Sharp, and also Cartwright, seem to overlook is that scientists *never* examine any real system in all its messiness. That simply is not — and has never been — the aim of science. It is however part of the task of philosophy of science to show how such abstract and general theories may be said to be (or not to be) about aspects of this complex reality, and yes, that is where studies of the internal structure of models of theories and the various relations into which they enter come in.

It seems that (for Cartwright) it is because of the simulacrum nature of models that bridging relations can only hold *ceteris paribus*. There is, however, at least, a structural error in her account as far as *ceteris paribus* conditions are concerned. The view that portrays these conditions as some kind of ingenious device cunningly designed by naive realists or staunch fundamentalists to "save theories from point-to-point testing" (Rueger & Sharp, 1996, p.103)[74] is rather misguided.

A theory holds *ceteris paribus* yes, but not in Cartwright's sense of the word. In my terms, to say that a theory "holds" means, per definition, that it holds (is "true") in a particular one of its models. To say now that it holds "*ceteris paribus*" adds *nothing* to simply saying that it is true. Moreover, there *is* nothing else about which it can be stipulated that it stays the same — *everything* is given in the model. *Ceteris paribus* clauses seem in Cartwright's terms to play a more and more important role the further away one moves from fundamental laws. In model-theoretic terms, however, they are necessary only at the level of scientific theories or linguistic systems, and become less and less active the closer to reality one moves. I claim — see also Chapter 2 — that they are *suspended* in their generality as soon as the theory in question is interpreted in specific models, rather than *activated*. The idealised nature of interpretative — and even empirical models — is not the result of specific *ceteris paribus* clauses, but indeed simply true to the nature of science. No real system can ever be examined, represented, explained, or described in its full complexity. That is simply not science's function.

In this sense Rueger and Sharp (1996, p.95) refer to the problem she has with the covering law account of explanation as the "unsoundness argument" and set it out as follows:

> If ... phenomenological laws could be *soundly* derived from more fundamental laws as the traditional [covering law] view would have it, then any successful comparison of the phenomenological consequences of the theory with the observations would count unproblematically as inductive support for the theory. Confirmation would flow upwards from the phenomenological level to the fundamental level. This flow, however, is staunched ... because phenomenological laws typically cannot be soundly deduced from more fundamental theories. To derive the former we usually need assumptions [*ceteris paribus* clauses] which are either false (distorted representations of the situation of application) or which contradict the fundamental laws themselves. Inductive support

cannot, therefore, be transmitted.

Claiming that phenomenological laws cannot "typically" be deduced from fundamental ones, is perhaps jumping the gun a bit. Is it not the case that Kepler's laws can be deduced from Newton's in a very sound way? Moreover the *ceteris paribus* clauses and other additional assumptions needed to validate the fundamental laws are suspended when models are constructed of some theory — as remarked above — and thus these clauses become more and more concretely realised as they set the boundaries for the truth of the theory, i.e. the clauses themselves (e.g. "no other forces act differentially on components of the system") become realised, i.e. true in the relevant models. Thus, model-theoretically, it is rather unclear how they can be understood to "contradict" the fundamental laws themselves (which are also true in these models).[75] These *ceteris paribus* conditions or clauses will usually be incorporated into the formulation of the law explicitly (as when stating that Hooke's law holds *as long as the elastic limit has not been exceeded*), or else implicitly and tacitly by common understanding.

A last remark on saving the phenomena: back to Cartwright's claim that laws that explain are not necessarily true. Rueger and Sharp again (1996, p.96):

> There is thus a trade-off between a theory's explanatory power and its (potential) truth: the more efficient a theory is in explaining or organising a large variety of different phenomena, the less can it be true or state the facts.

As Cartwright (1983, pp.72,73) herself has been stressing since *How the laws of physics lie* (1983),

> If we state the fundamental laws as laws about what happens when only a single cause is at work, then we can suppose the law to provide a true description. The problem arises when we try to take the law and use it to explain the very different things which happen when several causes are at work. This is the point of 'The truth doesn't explain much'. There is no difficulty in writing down laws which we suppose to be true: '*If* there are no charges, no nuclear forces, ... *then* the force between two masses of size m and m' separated by distance r is Gmm'/r^2'. We count this law true — what it says will happen, does happen — or at least happens to within a good approximation. But this law does not explain much. It is irrelevant to cases where there are electric and nuclear forces at work.

Well, yes, of course, *if one believes in an absolute notion of truth*, and if one believes that this absolute truth in science is *about specific individual situations* in their uniqueness. It is, I believe however, possible to speak only of theories being *true-in-some-model* and not of theories being *true qua nothing else*, i.e. absolutely or "universally" true. Cartwright (1986) does point out the fact that "truth" does not mean "universal", as mentioned before, but her worries in terms of combined laws continue.

I shall address the latter here by considering again her (related) worries concerning her interpretation of *ceteris paribus* clauses or conditions. She views these clauses as playing an important role in the explanatory power of the fundamental laws, in the sense that they determine what kinds of explanation are permissible because they lay down or record, in a sense, the nature of the abstractions from real situations made by the theory and its fundamental laws. It seems then that in this sense the conditions laid down by these clauses also determine the nature of the models of the theory in question in their function as part of the concretising mechanisms of science, such that they "adapt" contexts to "fit" the laws explaining the behaviour of the objects found within that particular context.

In model-theoretic terms the role — if any — of *ceteris paribus* clauses is somewhat different — as I have already pointed out. First, they are only of importance — if at all — as part of the linguistic expression of some theory. Cartwright's reconstruction of Newton's gravitational law (see above quote) — i.e. "If there are no ..." is simply wrong. The law states rather that "The gravitational force between two masses ..." *without exception*. The law is still absolutely and totally relevant when there are (also) other forces present! What she implies with her reconstruction of the law is simply not the case. The gravitational force is still there in cases where electric and nuclear forces are at work. We take the (vector) sum of all the forces on a particular body to see how it will behave. We *do* have theories and models about how different causal factors *combine* when acting on the same system in reality. A simple example is given by the vector addition of speeds, accelerations, and forces. The only kind of reconstruction of Newton's gravitational law that mentions other forces and factors should then simply be something like this: "*Even* if there are electric charges, nuclear forces, the sun shining on them, rain falling on them, ..., then still, everywhere, *under all possible circumstances*; the *gravitational force* between two masses ...".

A word now on Cartwright's abhorrence of the generality of fundamental laws and the fact that such laws say "nothing" about what "actually" happens. Feynman (1965, p.14) describes Newton's universal law of gravitation as "the greatest generalisation of the human mind" and Cartwright uses this notion of the generality of laws to illustrate her rejection there-of. Both Newton's law of gravitation and Coulomb's law (to which Cartwright also refers) say that "a force will be exerted, which is not the same as saying what will happen at all" (Spurrett, 1999). Cartwright (1980, p.77) writes:

> It is not the case that for any two bodies the force between them is given by the law of gravitation. Some bodies are charged bodies, and the force between them is not Gmm'/r^2. Rather it is some resultant of this force which Feynman refers to.

Cartwright's worry is that

> In the simplest case, the consequences that the laws prescribe must be exactly the same in interaction as the consequences that would obtain if the law were operating alone. But then, what the law states cannot literally be true, for the consequences that actually occur if it acted alone are not the consequences that would occur when it acts in combination (Cartwright, 1980, p.83; 1983, p.72).

Again, from my metaphysical minimalist stance, solving this problem should be done, I think, by looking at the nature of the relevant generalisations or laws and their combination, and not at the nature of what it is that they state.

Creary (1981) offers a criticism of Cartwright's views on the issue of parts and sums that rescues the facticity of fundamental laws in a way that is different from mine and that I shall not go into now. I shall end my discussion of this particular issue by noting — in David Spurrett's (1999) terms (which goes into far more detail than I can) — that Nagel has actually already given us the way to answer these kinds of questions. Cartwright — unsurprisingly enough — does not accept vector addition as a way to

132   CHAPTER 4

address her above worries concerning parts and sums, since in such cases, the components are not part of the resultant. Nagel however showed that we have to look at the specific interpretation of the notion of "sum" in these cases. He (Nagel, 1963, p.142) writes:

> ... no antimony arises from the supposition that, on the one hand, the effect which each component force would produce were it to act alone does not exist, while on the other hand, the actual effects produced by the joint action of the components is the resultant of their partial effects.

Nagel interprets the notion of sum of a certain kind of things such that it "... will not necessarily have the constituents as literal parts ... but will more generally be determined by some function" (Spurrett, 1999).

I shall use David Spurrett's (ibid.) explanation here since it captures my thoughts far briefer than I can express them:

> What function exactly applies in any type of case will be an empirical question on Nagel's view [as in mine], and he makes clear that the history of science teaches us to expect some variety: not all addition is linear, not all retains the added elements as proper parts, etc..

Spurrett (ibid.) continues:

> ... forces have direction and magnitude, and can thus unproblematically be represented by vectors. The standard operations for adding vectors work perfectly well in at least some cases where the vectors are forces: Millikan's experiments to determine the charge of a single electron, ..., Coulomb's own experiments on the force due to charge, the paths of projectiles, ... But all of these are cases which fall in areas where Cartwright urges us to regard the laws as false, since they do not meet the requirements of the facticity view. That is to say, ... there is more than good enough reason to regard the scope of laws as extending some of the way beyond what Cartwright allows. ... To say that the law cannot apply because we cannot do the sums, or because other factors will make the outcome different to what it would be if the law acted alone is to get the epistemological cart before the ontological horse.

Moreover, returning to our discussion of Cartwright's interpretation of *ceteris paribus* clauses, these clauses, where necessary, form part of (a complete formulation of) the law and are not even really extraneous conditions. A scientific theory makes statements concerning the nature and behaviour of a certain phenomenon, or a group of phenomena, in some real system(s). These statements are "sweeping" precisely because they have to cover *all* phenomena, in whatever context they may occur, that may exhibit the features of the ones described in the theory, and not only specific ones. In that sense, the formulation and application of any fundamental law is never *ceteris paribus*. Rather than saying "if all possible influencing factors not explicitly mentioned are absent or neutralised, then ...", a fundamental law will typically say "even if all possible other factors influence the system in all possible ways, then still ...". Of course, the complete formulations of many (fundamental and phenomenological) laws are conditional. Remember again Hooke's law: "If the elastic limit has not been exceeded, then ...".

The specific clauses that Cartwright has in mind, are thus much rather part of the content of the (conditional) law expressed by the theory in question, than conditions for the law's applicability. Ironically enough, if these clauses are conditions (in

Cartwright's or in the logical sense) it implies that they can be negated — i.e. not be satisfied — and the conditional formulation of the law can still apply. (It is a case of simple logical equivalence that $cp \to law \leftrightarrow \neg cp \lor law$). One reason why Cartwright sees *ceteris paribus* conditions as separate from the law and will probably not accept my incorporation of them into a conditional formulation of the law, is the following. Her *ceteris paribus* conditions may involve influencing factors (for the physical system under consideration) for which there are not even terms in the language of the theory. Look again at her reconstruction of Newton's law of gravity: it drags in *ceteris paribus* conditions involving electricity and nuclear forces, about which the language of Newton's theory cannot even talk. So it is impossible to formulate her reconstruction of the law as a conditional sentence in the language of Newton's theory.

What we have here is another manifestation of Cartwright's worries concerning scientifically and truthfully encompassing a system in all its limitless complexity and interrelatedness. Perhaps Cartwright misinterprets the reason for the fact that truth and universality are different concepts. She (Cartwright, 1997, p.167) claims that "To say the laws of physics are true *ceteris paribus*, is not to deny that they are true. They are just not entirely sovereign". Well, model-theoretically, theories are indeed not simply true, whether conditional, *ceteris paribus*, or not, but they are sovereign. Theories can only be *true in their models*, regardless of how many — if any — *ceteris paribus* clauses form part of their formulation. Moreover it is exactly because they are sovereign, in the sense of being formulated for all possible circumstances satisfying their terms, that they have any chance at all to be true (in their models).

Now, finally then, let us look again at Cartwright's distinction between universality and truth. She (Cartwright, 1989, p.162) claims that theories should not be taken as summaries of laws about observable entities, because theoretical entities are *not* needed to *explain* the behaviour of observable entities, but are rather necessary to *systematise* observable behaviour — as pointed out before. Very much in the constructivist tradition, she then goes on to stress that theories are never universally applicable, but their domain (the limit of their applicability) is determined by making use of the theory and its concepts themselves. This is interesting in the sense that, as far as the constructivists are concerned, it makes it impossible to ever move to a meta-level for any reason — like evaluating the scientific content of a theory. What Cartwright wants to show, I think, may also be along these lines, although she is, more specifically, aiming to show that truth and universality do not necessarily imply each other. This, of course, is entirely in line with a model-theoretic interpretation of these notions. The notion of universality, in model-theoretic terms, is not applicable when it comes to science, and the notion of truth, though still important, is ultimately based on the Tarskian notion of satisfaction and on context-dependent empirical adequacy.

All of this may be viewed as pointing to the necessity of the interpretative role models play in science. In an article entitled *The tool-box of nature* (1995), that she co-authored with T.Shomar and M.Suárez, Cartwright again claims that "... representations of phenomena must be *constructed* and theory is one of the many *tools* we use ..."(ibid., p.139). (My italics.) She (ibid., p.140) goes on:

> I want to urge that fundamental theory represents nothing and there is nothing for it to

> represent. There are only real things and the real ways they behave. And these are represented by models, models constructed with the aid of all the knowledge and technique and tricks and devices we have.

I have no quarrel with these remarks. That is exactly what I am trying to show, in the sense that I want to establish the fundamentally "constructed" nature of science. But, although in a sense I urge the "constructedness" of theories just as much as the "constructedness" of models, I view the role of theories and their abstract and general nature as a central part of what science *is*, while Cartwright more often than not sounds as if she would much rather do without theories, although she of course does acknowledge their organising features. And the reason for that is, I think, that, because of her antagonism towards fundamentalism, she focuses too much on the spurious theory-reality link at the cost of the construction of the theory-model-reality link.

To summarise, in a realist context such as mine, I see the role of theories within the whole representation process as more meaningful — and maybe more useful — than Cartwright sometimes seems to do. In my view, theories are absolutely essential to science, because they formulate the conceptual content of the models, make this content amenable to deduction and computation, and communicate this content to other scientists. To me models are constructs, that is (interpretative, i.e. mathematical) structures that are not primarily linguistic, while theories are primarily linguistic entities (sets of sentences of some appropriate language, that — among other things — describe models in which those sentences are true).[76]

Cartwright and company (ibid.) are advocating a rather "free" notion of models in the sense that (phenomenological) model construction should be viewed as (almost) independent — in method and aim — from the theory in question. In a sense, at least as far as empirical over-determination (Chapter 2) is concerned, it might seem as if this fits somehow into my framework too. The *semantic* link between theory and model in model-theoretic realism, however, *has* to remain. Establishing this link is not applicable or even really relevant in a non-statement framework, given non-statements supporters' disregard of scientific theories as linguistic entities (although most of them do not deny that models are *interpretations* in which a given theory is true).

Now that we have examined a model-theoretic realist account of science (Chapter 2), and discussed it in terms of both typical statement (Chapter 3) and non-statement (this chapter) accounts of science, in the next chapter, we shall refine certain of its main (particularly realist) characteristics.

NOTES: CHAPTER 4

[1] Sections from this chapter are published as Ruttkamp (1999a).

[2] See Suppes (1954, p.244).

[3] Suppes (quoted in Wójcicki in Humphreys, 1994, pp.148,149) writes: "The more I think about scientific practice and reflect on how to give an accurate account of the complicated processes that go into experimentation, the more I am persuaded that there are a large number of distinctions needed to describe experimentation thoroughly, especially as data are purified for quantitative, and even more statistical, analysis. It is a long way from running around the laboratory doing one thing and then another, to having a set of data as printout or on a computer screen ready for analysis. That process still needs much more thorough attention ... gruesome details of exactly how data are purified and selected for analysis, not to speak of details of how they are generated, which itself may involve, as equipment becomes increasingly complicated, many different independent tests of reliability and accuracy of equipment".

[4] Issues concerned with this stage of science are addressed, for instance, by Paul Galison in his book entitled *How experiments end* (1987), and the details have now been worked out to unbelievable depths in his follow-up *Image and logic* (1997).

[5] "The kind of co-ordinating definitions often described by philosophers have their place in popular philosophical expositions of theories, but in the actual practice of testing scientific theories a more elaborate and more sophisticated formal machinery for relating a theory to data is required" (Suppes in Morgenbesser, 1967, p.62).

[6] To be able to establish a representation theorem for a theory implies that it can be proved that there is a class of models of the theory such that every model of the theory is isomorphic to some member of this class. Suppes (1960, p.295) gives a few examples of such theorems, for instance, Cayley's theorem that every group is isomorphic to a group of transformations, and Stone's theorem that every Boolean algebra is isomorphic to a field of sets.

[7] The language used to formulate a particular theory is the same one used to formulate a particular reduct. Recall that a "reduct" in model-theoretic terms is created by leaving out in the language and its interpretations some of the relations and functions originally contained in these entities. This kind of structure thus has the same domain as the model in question but contains only the extensions of the empirical predicates of the model.

[8] Suppes (1988b, p.254) claims that one of the most important and valuable uses of representation theorems in philosophy of science is that they help to increase (scientific) understanding of the represented object.

[9] If a representation theorem is found for one science in terms of a second, the first has been (formally) reduced to the second — e.g. Adams (1959) was the first to give a rigorous proof of reducing rigid body mechanics to particle mechanics.

[10] The "type" of a model is determined by the individual constant symbols, as well as by the relation and function symbols of the axiomatic calculus of the theory in question.

[11] See Mayo (1997) and Galison (1997) for recent discussion of this process.

[12] See Ruttkamp (1997b), and the discussion related to these issues in Chapter 2.

[13] Sneed (1983, p.350) claims that structuralism "... is essentially a view about the logical form of the claims of empirical theories and the nature of the predicates that are used to make these claims". (The notion of 'predicates' is taken in the usual set-theoretic sense of characterising the type or species of sets of structures.)

[14] For formalised theories the entire (meta-) mathematical apparatus for studying theories and their models becomes available to the philosopher of science. One example of the tremendous usefulness of this approach is the study of verisimilitude.

[15] Both Stegmüller and Sneed formulated reconstructions of parts of Kuhn's theory, touching on the role of the scientific community in the development of scientific theories (Stegmüller, 1976; Sneed, 1976). Sneed claims that "... in order to make sense of what Prof Kuhn was telling us about scientific activity ... we found it convenient to ... employ a concept of scientific theory somewhat different from that commonly used by philosophers of science-in-general" (Sneed, 1976, p.119). He is referring here to the adoption of the "non-statement view" by him and Stegmüller (and their followers). (Of course, Kuhn's work was not the only

motivation in this regard, as Sneed himself acknowledges.) As already pointed out in the above, Sneed and Stegmüller both reconstructed parts of Kuhn's philosophy of science, focusing especially on the notion of a member of some community "holding" a particular theory, which implies a concentration on the differences between normal and revolutionary science. Briefly, Sneed (1976, p.120) defines "normal change" as change in the body of empirical claims of a theory, while "revolutionary change" consists in the changing of theories themselves. These notions can be very successfully treated by the structuralist programme, and Stegmüller (1973) specifically showed that the relation of reduction can be of significant use in depicting the notion of scientific progress. Kuhn (1976) wrote an article in reaction entitled "Theory change as structure change: Comments on the Sneed formalism".

[16] Recall that in general, a model is a structure (interpretation) of the form $<A_1, ..., A_m, R_1, ..., R_n>$ where the A_i are the "basic sets" or domains of the model (the ontology of the theory); and the R_j are relations on the A_i. Remember also that — at least for a language with a sound set of rules — satisfaction of the axioms implies satisfaction of the theory, for any interpretation.

[17] We all know that a theory usually has many different models, but they all have one thing in common, which Balzer, Moulines and Sneed (1987, p.3) identify as the same structure, while I emphasise also the fact that they are all models of the same (linguistically expressed) theory. A theory offers one formulation which binds together all these models (e.g. think of a theory as a set of field equations). That is why the model-theoretic approach is the one that I choose. This approach offers the possibility to focus on the linguistic nature of the theory *as well as* on its different models. Be that as it may, I do agree with Balzer, Moulines, and Sneed (1987, p.3) that what is meant by models sharing the same structure, is that they all share the same conceptual framework (i.e. in my terms, they all have the same logical type or signature) and they all satisfy the same laws (theory).

[18] Sometimes also referred to as "conceptual determinations".

[19] "... this distinction may be understood as the model-theoretic explication of the distinction between the 'analytic' and the 'synthetic' components *within a particular theory*" (Moulines in Schurz & Dorn, 1991, p.318). Or perhaps, in my terms, this may be viewed as the distinction between the themata and related context-specific factors, in so far as these co-determine the logical type (signature) of structures, and the linguistic formulation of the empirical claims suggested by the interpretation of the empirical data in question.

[20] See Balzer, Moulines, Sneed (1987, pp.19,20).

[21] Briefly, the intensional description of I is a description in terms of the properties of I, while the corresponding extension of the set I denotes the elements of I — i.e. which elements of I have these (intensional) properties.

[22] In my terms, the elements of I would be *representations* of systems of the "real things".

[23] Without a distinction between theoretical and non-theoretical terms, structuralists simply say that a particular intended application is an element of M_P. If such a distinction is made, they say that a particular intended application belongs to the class of partial potential models, M_{PP}, which is formally derivable from M_P.

[24] Or of most depictions of models of theories in formal terms, for that matter.

[25] Approximation has been left out of this discussion, simply because the inclusion of approximate relationships will only complexify matters needlessly, since this discussion is meant as a brief introduction into the structuralist programme. I think it suffices for my purposes to make it clear that within the structuralist programme all approximate relations can be defined formally and are definitely taken into account in their reconstructions of empirical theories.

[26] See Moulines in Shurz & Dorn (1991, p.324).

[27] This simply means that the set I has "a life of its own" (ibid.), in the sense that its endurance is not dependent on the endurance of its members. The issue is especially complex, because the "life" or nature of the class I that endures through time (or history) depends on the nature of the *scientific community* to which it is linked, which, in its turn, is also a "genidentical" entity.

[28] Note, however that although the set of intended applications cannot be depicted in purely formal terms, constraints and inter-theoretical links can be formulated formally by using structural descriptions of models of the theory.

[29] "Intended" here refers not to the formulation stages of theory development as I have set it out in Chapter 2, but rather Balzer *et al.*. want to focus on the particular application (interpretation in my terms) of a specific theory to a certain real system or "range of phenomena".

[30] Note again that Sneed and company take a theory element as the core of a theory plus its range of intended applications. (See Sneed (1976).)

[31] See Sneed (1976).

[32] "The intuitive idea is that a distinction may be drawn between what is ruled out by the structure of the theory's models M and what is ruled out by restriction on the way that structure is applied 'across' a number of different applications C" (Sneed, 1976, p.124) — "Local applications [of theory T] may overlap in space and time, they may influence each other (even if they are separated in time and space), certain properties of T's objects may remain the same if the objects are transferred from one application to another one. Any connection of this sort will be captured by what we call *constraints*" (Balzer, Moulines, Sneed, 1987, p.41).

[33] See Balzer, Moulines, Sneed (1987, pp.46ff., Sections II.2.3 and II.4) for more detail.

[34] The sections of the structuralist programme dealing with these issues are very technical — see Balzer, Moulines, and Sneed (1987, pp.57ff., Section II.3.2 ; pp.73ff., Section II.3.4) for more detail.

[35] See also Sneed (1976), and Balzer, Moulines, Sneed (1987).

[36] This is done (ibid.) in such a way that the whole array of theoretical components satisfies the constraints C.

[37] See Balzer, Moulines, and Sneed (1987, pp.57ff).

[38] All of the above is of course set out in idealised terms since it is not, in this context, taken into account that the empirical claim associated with a particular theory element will always — according to the structuralists — be only approximately true. What is relevant here in this connection is not the overwhelming literature on the technical aspects of the question of approximate truth, but rather, and much more simply, investigating exactly what the structuralists envisage the theory core's function to be in all of this. The briefest answer is, obviously, that the theory core identifies the theory content. More precisely, the theory core defines a set of possible situations or "ways things could be" (Sneed in Humphreys (1994, p.195)), called content(K). This notion of "ways things could be" again link up with the need I see for preferential analyses of groups of empirically equivalent models, discussed in Chapter 2. I shall not go into any more detail as far as these issues are concerned though, given the scope of this chapter.

[39] See Beth (1949), (1961), and Van Fraassen (1970).

[40] See Suppe (1967), (1973), and (1989).

[41] Van Fraassen (1970) points out that Wilfred Sellars has since the late fifties been arguing for precisely such a meaning structure for the language of science. See Sellars (1957, pp.225 - 308), and (1963, Chapters 4, 10 and 11).

[42] See Chapter 2 again.

[43] The context-dependency of mathematical models does however also enter in their view of scientific theories in so far as they — especially Suppe (1973, pp.151ff.) — discuss the "extra-theoretical factors" determining for instance the experimental design of a theory. These factors include "regularities" (ibid.) such as other theories, laws or known regularities about the phenomena in question.

[44] Giere (1983, p.271) explains that a physical system "... is defined by a set of state variables and system laws that specify the physically possible states of the system and perhaps also its possible evolutions". Giere offers the example of classical thermodynamics which may be understood as defining an ideal gas in terms of three variables: pressure, volume, and temperature, and then specifying that these are related by the law $PV = KT$. (See also Suppe, 1973, p.132.)

[45] Van Fraassen (1970, pp.130-132) discusses three types of law:
- *Laws of succession* are relations of succession indicating the various sequences of states various physical systems will assume over time. These relations are such that the sequences may be deterministic or statistically determined, continuous or discrete. These laws (as far as they are non-statistical) thus select the physically possible trajectories in a particular state-space.
- *Laws of coexistence* are equivalence relations indicating which states are equivalent to which others, if the associated law is deterministic. If it is statistical it indicates which states are equally probable,

i.e. it selects the physically possible subsets of the given state-space.
- *Laws of interaction* (either deterministic or statistical) determine which states result from the interaction between various systems. These laws are combinations of the first two kinds of law.

[46] See Beth, E., "Carnap's views on the advantages of constructed systems over natural languages in the philosophy of science" (especially pp.479-480) in Schlipp, P (ed). 1963. *The philosophy of Rudolf Carnap*.

[47] "Suppose that for a given kind of physical system, the theory specifies a set E of elementary statements, state-space H, and satisfaction function h. We call the triple $L = <E,H,h>$ a *semi-interpreted language*" (Van Fraassen, 1970, p.335).

[48] Especially non-relativistic physical theories typically use mathematical models to represent the behaviour of a certain kind of physical system — Van Fraassen (1970:328) gives as examples the use of Hilbert space in quantum mechanics, and the use of Euclidean $2n$-space [sic] as phase-space for n particles in classical mechanics. (He probably means "... for a system with $\underline{n}$ degrees of freedom ...". For $\underline{n}$ particles $6n$-space is needed!)

[49] This is much in line with my idea of an interpretative model of a theory and relations linking it to some real system via some empirical model offering a "snap shot" view of the real system at a specific time.

[50] "For each elementary statement U there is a region $h(U)$ of the state-spaces H such that U is true if and only if the system's actual state is represented by an element of $h(U)$. (We also say that these elements *satisfy* U ...)" (Van Fraassen, 1970, p.329).

[51] This notion of a "satisfaction function" characterises the kind of relation I see involved in determining the possible isomorphic embeddings of empirical models into interpretative models of some theory.

[52] See Van Fraassen (1970, p.337).

[53] "For each elementary statement U there is a region $h(U)$ of the state-spaces H such that U is true if and only if the system's actual state is represented by an element of $h(U)$. (We also say that these elements *satisfy* U ...)" (Van Fraassen, 1970, p.329).

[54] A science like physics cannot exist at all if it were not to use rich mathematical structures and the deductive methodology of mathematics and restrict itself to empirical models only. For example, no physical measuring process can ever produce an irrational number as result, although interpreting some of these results necessarily assumes the existence of such numbers.

[55] See Niiniluoto (1999, pp.115ff.) for a discussion of the anti-realist aspects of constructive empiricism.

[56] This is yet another appropriate moment to refer you at least to Mayo (1997) and Galison (1997).

[57] "For our purposes, a scientific theory has two components. One is a family of [theoretical] models ... . The second is a set of theoretical hypotheses that pick out things in the real world that may fit one or another of the models in the family" (Giere, 1991, p.29).

[58] See my discussion of this issue in Chapter 5.

[59] See Wartofski (1979, p.19).

[60] He (Giere, 1984:13) gives as example of a theoretical hypothesis the statement that "The positions and velocities of the earth and moon in the earth-moon system are very close to those of a two-body particle Newtonian model (with specified initial conditions)".

[61] Another example of a notion of approximation by degree, as noted by Giere (1985:80), can be found in Beth's state-space approach in terms of the value r of a magnitude m at a time t in a given state-space. This notion also fits both my and Suppes's approaches to these "empirical" relations.

[62] And so they should be, given that they form part of human scientific knowledge which, from the beginning, simply *is* based on activities of abstraction, because that simply is how we humans *know* anything.

[63] We have already agreed that fundamental laws are indeed too simple and abstract to be *directly* about any aspect of reality.

[64] She also remarks in Cushing, Delaney, and Gutting (1984, p.135) that "... abstractness and scientific realism are two different issues, and not all varieties of abstractness bear equally on questions of descriptive completeness, accuracy, and truth. This is so with our notion [of abstraction], where notions become more and more abstract as less and less explanatory information about them is given".

[65] Both Cartwright and I view these models as idealisations, although we differ about the implications of the ideal nature of these models for the process of science, as will be discussed below.

[66] Cartwright (1983, Chapter 8).

[67] Cartwright's "phenomenological laws" remind one very much of Suppes's (1989) models of data.

[68] See also Chalmers (1987, p.82) for confirmation of this interpretation.

[69] See also Chapters 2 and 5.

[70] Well, this is true of fundamental laws too — Newton says openly he offers no hypotheses concerning the *reasons why* his laws of gravitation are true.

[71] See also my notes concerning Newton's mechanics offering an explanation of Kepler's laws in Chapter 2.

[72] Cartwright claims (1983, p.150) a model realistic in this second sense to be in need of more bridge principles than one realistic in the first sense. The best explanation for this is, I think, the fact that she sees the mathematical representation as being closer to — or perhaps mainly identical to — the theory.

[73] As she explains (Cartwright, 1983, pp.152-154): "The second definition of 'simulacrum' in the *Oxford English Dictionary* says that a simulacrum is 'something having merely the form or appearance of a certain thing, without possessing its substance or proper qualities'".

[74] Cartwright illustrates this accusation with the following remarks: "Not all radiometers that meet Maxwell's two descriptions have the distribution function Maxwell writes down; most have many other relevant features besides. This will probably continue to be true no matter how many further corrections we add. In general ... the bridge law between the medium of a radiometer and a proposed distribution can hold only *ceteris paribus*" (Cartwright, 1983, p.155).

[75] There are cases in which we believe phenomenological laws to be soundly deducible from a certain set of fundamental laws, but find that the actual deduction is extremely difficult. These cases, however, do not prove in any way either that phenomenological laws "typically" cannot be deduced from fundamental ones, or that the "all things being equal" and additional assumptions needed in such deductions may be found to "contradict" the original (set of) fundamental law(s).

[76] The main means of communication in science still is language (together with diagrams, physical models, demonstrations, films, and so on).

CHAPTER FIVE

A MODEL-THEORETIC REALISM[1]

1. INTRODUCTION

The realism issue threads through all of the above, whether in the statement account context from Feigl's form of realism to Carnap's "neutralism", or in the non-statement context where mostly we find various forms of anti-realism, except when the notion of truth is addressed in terms of truthlikeness, such as in Niiniluoto's case. No examination of the nature of scientific theories can be complete without addressing the relations between these theories and reality. In a model-theoretic model of science such as mine the basic ontological assumption made is that science is about something that exists independently of it. This ontological assumption has however as little metaphysical content as is possible. Claiming that reality exists "outside" of human practice neither means that reality is unknowable nor, at the other extreme of the scale, that science simply mirrors it.

To determine whether science is about an independent reality or not, I believe one has to examine both the actions necessary for producing scientific knowledge and those aiming to apply such knowledge. From the previous chapters it should be obvious that I argue that at least some of the terms of a given scientific theory possibly refer to entities and relations of some system in reality if reference has been assigned to these terms via some empirical reduct of some interpretative model of the theory. In other words, a scientific theory can be said to be "about" the potential of entities in some system of reality to give reference to some terms or objects in this way. This potential however, is independent of the existence and knowledge of humans. It is the actualisation or realisation of this potential that requires human action. (See also Chapter 2.)

As has been mentioned before, any successful model of a theory is guided by contingent conditions which are context-specifically constructed. The range or content of these models thus cannot be established *a priori*, but has to be investigated *a posteriori* by means of historical studies of the development of science. We can however at least say (*a priori*) that formally, the abstract linguistic character of scientific theories allows the possible existence of (many different) models constructed to interpret them; and that these models (with the help of their empirical reducts) have the ability to realise the potential theories have to refer to real objects or relations and also the potential objects and relations in real systems have for being the referents of terms in scientific theories. The human-independent potential of reality and its contingent context-specific realisations are thus conceptually speaking (in terms of human input) fundamentally different, even if they are epistemologically (and methodologically) related. "Reality" and "knowledge of reality" — i.e. ontology and epistemology in their general senses — are separate and not to be confused. The historical reality of theories and models is dependent on the actions of humans, while the reality of the ongoing potential of objects and relations in real systems for being the referents of terms in these theories and models is not. The need for acknowledging the conceptual stratification of the conceptual development of science should now be even more obvious, since at the level of the theory alone, all that is apparent is the multi-interpretability of the theory and some possibility of it finally referring to a variety of systems in reality.

In *The Scientific Image* (1980), Bas van Fraassen remarks that realists believe scientific theories offer "a faithful replica, in all detail, of our world". I am arguing (in contrast to Van Fraassen's somewhat straw man-realist) for a view of realism that is more sophisticated in the sense that the intricacies of the various semantic relations between reality — in terms of real systems — and scientific theories are taken into account and it is acknowledged that absolute universal statements about these relations fit as uncomfortably within a scientific realist framework as in any other framework.[2] The identification (discovery) of particular instances of the general terms of theories becomes possible only through interpretations of these general terms, such as mass, momentum, and so on, in context-particular models of the theories.

As acknowledged above, the possibility of discrepancies between theories, models and actual systems in reality has to be acknowledged and as Nancy Cartwright, Ronald Giere and their allies rightly advocate, the need for particular conditions to make "real" sense of general theories has to be accepted too. But this neither leads necessarily, for instance, to Van Fraassen's notion of empirical adequacy concerning only models and phenomena, nor to typical scientific realist claims that real systems are approximately captured by models of theories, nor to other problems such as the negative implications (for realists in my sense) of Putnam's model-theoretic paradox.

In contrast to all of these, I claim that there may be a set of models of some theory which fits some systems of reality and add that it is via historical studies concentrating on the context-specific models already constructed to interpret the relevant theories — and perhaps by constructing some new ones — that this can be determined. It is this inherent feature of science — that is the multi-interpretability of its theories — which should be emphasised if a sophisticated realist interpretation of the conceptual development of science is offered. And the multi-interpretability of scientific theories is the realist counter to the abstract nature of science. In this sense, it seems natural that it is so difficult to describe the relations between science and reality in simple unique terms.

Somehow Ronald Giere's (in Churchland & Hooker, 1985, p.82) remark that man is not the measure of all things but rather the measurer, seems very fitting in this sense. The structure of science's development hangs necessarily together with its theoretical nature. The content of verification procedures for the content of science however are never given beforehand by fixed rules, but are a result of the context-specific actions and constructions of human scientists. Notions like "constructed reality" and "real science" should therefore not be viewed as representing opposing interpretations of science, but should rather be accepted as the two poles of one very complex time-bound relationship.

2. REALITY AND SCIENCE

Model-theoretically speaking science studies systems in reality. This refers to the abstracting simplifying nature of science. As already often pointed out in the above, no-one, not even scientists, can study reality in all its fullness at once. Not only do scientists focus on some particular system of phenomena in reality, but also they aim to "adjust" that system in such a way that they can focus only on certain of its features. The kind of abstraction that Cartwright talks about and that I have discussed in a model-theoretic context in Chapter 2 is a necessary and sufficient condition for scientific knowledge. If certain abstractions are made from the richness of experiences that reality has to offer, scientific knowledge of the real system in question becomes possible. And, *vice versa*, if some knowledge claim is offered as part of science, the nature of that claim will (relative to the complexity of the universe) be simple and it will be about a sufficiently abstract version of some real system (even if that system is the cosmos!).

Reality is not unknowable, but rather only knowable in a certain way. I am aware that this sounds particularly Kantian, but that is, in a sense, my purpose. The basis of a model-theoretic methodology is Kantian in the qualified sense that it implies both that we can only know reality through science, and that scientific knowledge can only be achieved through certain abstractive actions. However, the Kantian *Ding-an-sich* is the reality we study via science. There is no other "underlying mechanism" or anything else that is somehow so fundamental to the ontology of reality that we cannot know it. Knowing through abstraction is knowing. There is no other kind of knowing, scientifically speaking. And, moreover, this kind of knowing is adequate in the sense that it does indeed allow us to study, discover, and utilise knowledge about reality.

Any kind of realism that decides how reality has to be in order for science to be possible — such as that of Bhaskar (1978) — is too metaphysical for my taste. I have been arguing all along — and pointed out many times — that a model-theoretic realism is one that focuses more on science than it does on reality. Connecting an ontology of science to an ontology of reality however often seems to be strangely lurking behind many philosophic accounts of science. It seems as if Cartwright (1989), for one, has also been seduced. Her hierarchy of "capacities" somehow seems to be offered as a kind of mirror image of her hierarchy of scientific laws — fundamental laws are capacity claims, phenomenological laws depict causal relations in reality, and the experimental stage of science focuses on "singular causings". Roy Bhaskar ((1979), (1981)) also seems to have been caught, albeit perhaps in a different, but not necessarily lesser, way. In Bhaskar's terms the mechanisms of reality form part of the lowest — ontologically deepest — level of reality and their scientific counterparts are causal laws. The actual events produced by these mechanisms still form part of the "intransitive" (real) dimension of science but are conducive to the identification of constant conjunctions or patterns of events at the "transitive" (constructed) dimension of science.

From the above — as has already been implied in the introduction to Chapter 2 — it is obvious that I propose a distinction between the ontology of reality or of the physical world, on the one hand, and the ontology of science (which includes an epistemology of scientific knowledge) on the other. The reasoning behind disclaiming that such a distinction is possible seems to me as mainly based on the notion of science as a social construction. This is another of many contemporary echo's of Bachelard's (1934) point that it is difficult to see how science can be said to be about a human-independent reality if our notion of reality seems to be so very dependent on human action. One of the problems here, I think, is a certain vagueness of terminology.

In trying to clarify this confusion, let us first make it clear that in model-theoretic terms science is indeed also an individual and social construction. Science is "transitive" in Bhaskar's sense as against the "intransitivity" of reality. Scientific knowledge can change and is contingent on the actions of scientists formulating it. However the notion of "reality" is sometimes used as a group noun to refer to the immediate "stuff" or results of scientific knowledge. That is, the idealised pictures of the "external" world that science offers us are somehow seen as constructing a reality about which science is. Since these "images" are also still changeable because they are part of science, it is then perhaps concluded that reality too is changeable and therefore that a scientific epistemology is necessarily linked to an ontology of this (scientific) reality.[3] The "immediate" pictures that models of scientific theories offer of some system in reality are however just *that* — i.e. representations of reality. The reality that is independent of science, in the sense that it exists regardless of whether its processes have already been "discovered", "explained", or "described" by science, is perhaps better denoted by the — somewhat outdated — term "Nature".

Reality in *this* sense is not *dependent* on human actions at all. It might indeed be a complex system of "mechanisms" (or "capacities") the processes of which continue now as they have done through the ages. And, it is the ontology of reality in this sense that *is* separate from scientific epistemological factors. The "reality" of science is an idealised version of this "Nature" and consists of already established theories and the various actions of scientists as well as the results of these actions. This "reality" is a social construction and this surely is the reality the more moderate constructivists claim to be constructing, in the sense that science cannot be studied without taking these matters — i.e. human activities, their context-dependent motivation driving the scientific process, and also already established theories — into account.

The ontology of this "scientific reality" and the epistemology of its generator — i.e. science — can indeed in a certain sense not be separated. In a model-theoretic account of science this becomes even more clear. The models that offer us these pictures are an integral part of the process of science. The set of conceptual models referred to as the "intended" models of scientific theories in the above, are shaped by the already established pictures of this kind. And, moreover, at the interpretative stage of science the justification for, and evaluation of, scientific theories are offered and carried out within the context of this "scientific reality". The point is however, that in the final instance, although science is "social" and "constructivist" it is not a reflexive enterprise in the usual social constructivist meaning of the word. Science is not about something that it has constructed itself and that is inherently of the same nature as science itself. Science is about "Nature" and about discovering the intricacies of the mechanisms according to which "Nature" operates. In *this* sense, as pointed out above, ontology — in the sense of the ontology of that *about which* science is — and epistemology have to be separated.

The "game" of post-Kuhnian philosophy of science has one trick that has to be learnt if the realist quest is to be salvaged. This is the trick of keeping constant one of the changing factors at issue (i.e. "Nature" in this case) while acknowledging the variability of all the (transitive) others involved. Bhaskar (1978) claims that the relationship between science and reality seems problematic only if one either accepts the social character of science, but denies that its object of study is independent of all social activity (the so-called "epistemic fallacy"), or if one accepts the independence of reality, but denies the social nature of science (the so-called "ontic fallacy").[4] A model-theoretic realism is not guilty of either of these fallacies. Bhaskar (1989) depicts the basis of a valid realist philosophy of science in terms of a choice between either an epistemological or an ontological relativism. He chooses an epistemological relativism and refers (ibid., p.57) to it as the "... correct thesis of epistemic relativity, which asserts that all beliefs are socially produced, so that all knowledge is transient, and neither truth values nor criteria for rationality exist outside historical time ...". To ontological relativism he (ibid.) refers as the "... incorrect thesis of judgmental relativism, which asserts that all beliefs are equally valid, in the sense that there can be no (rational) grounds for preferring one to the other".

Making the choice — as a model-theorist would also do — for Bhaskar's epistemological relativism, and thus acknowledging that truth criteria as well as criteria for (scientific) rationalism are part of the philosophy of science, does not mean that any criterium goes, but rather the opposite. The construction of these kinds of scientific criteria is, model-theoretically speaking, undeniably a function of the interpretative <u>and</u> empirical models of scientific theories. *Validating* them is the function of the various semantic relations that exist between reality as "Nature" and these models.

3. A MODIFIED IMAGE OF SCIENCE

Around 1960 philosophers of science found themselves in somewhat of a crisis. The logical empiricist inspired way of thinking about scientific knowledge as being the "crowning achievement" of human reason suddenly seemed horribly empty. It began to seem less than clear what exactly it is that determines the line of research science should pursue at a given time, and, consequently, to determine which scientific theories are finally getting at "the truth". Philosophers of science became obsessed with questions like "What do we really know?", "What should we believe?", "What is evidence?", "What are good reasons?", "Is science as rational as people used to think?".

These questions lead to serious uncertainty about the nature of reality: "What is the world?", "What kinds of things are in it?", "What is 'true' of them?", "What is 'truth' in this sense?", "Are the entities postulated by theoretical physics real, or only constructs of scientists' minds for organising experiments, or worse, simply figments of their fertile imaginations that come in handy when they start formulating theories?".

Philosophers of science have already been divided over the answers to these questions for centuries. During the sixties — and ever since — the debate between realists and anti-realists has become far more intense though because of the strong historical and social flavour of the specific crisis they found themselves in. The opening phrases of Kuhn's *The structure of scientific revolutions* (1970) decided the context of this debate: "History, if viewed as a repository for more than anecdote or chronology, could produce a decisive transformation in the image of science by which we are now possessed" (ibid., p.1). (See Hacking (1983).)

Philosophical debates about science before 1960 range over the following image of science that Kuhn's work rejected. Participants of these debates argued (to varying degrees of fanaticism) (ibid., pp.5,6) — among other things — that

- there is a clear division between *observation* and *theory*;
- the growth of scientific knowledge is *cumulative*;
- science has a pretty strict *deductive structure*;
- scientific terminology ought to be as *precise* as possible;
- there is a distinct and fundamental difference between the *context of justification* and the *context of discovery*.

Not only did Kuhn not accept any of the above, but he also stressed that, above all, science is *essentially historical*.

My arguments thus far imply that scientific actions — and their authority — are not about Nature in the traditional confirmational sense of satisfying some set of a-temporal methodological rules and offering a body of neutral pure objective data about reality. But, neither do the entire scientific enterprise and its products offer simply sets of (false) context-specific data. Rather, science is about "Nature" in the sense that it is a system of knowledge claims that operates according to a set of rules that results in a body of contingent[5] data about systems in reality that offers us "snapshots" of "Nature".

Let us look again at the features of the pre-Kuhnian image of science Hacking identifies, to see how — if at all — they are depicted by a realist model-theoretic account of science. First let us consider the distinction between observation and theory. I think it is clear by now that a model-theoretic realism implies that observational and experimental actions and actions of formulating and applying models and theories take place at different stages or levels of the scientific process, although they constantly complement each other. Abstractions based on established and new observations play a major role at the beginning of the process when the intended model of some future theory is being constructed. Observational results also direct — together with the thematic preferences and other specific factors guiding the interpretations of scientists at this stage — the transition from the intended model to the theory in the sense of the *interactive* conceptual movement between this model and the selected aspect of reality. In formulating the theory — and thus fixing its class of possible models — the intended model, but also the variety of features in reality that are to be dealt with by the theory and its models, must be taken into account. During the interpretative stage of science observation again plays a definite role in the form of experimental data embodied in empirical models embedded into interpretative models of the theory in question. Throughout all of these stages it has to be noted though that established theoretical frameworks — especially those concerning the interpretation of experiments and the use of instruments — are never separate from observational activities.

Observation thus permeates the whole process of science, and moreover, the old fixed distinction between "observational" and "theoretical" terms dissolves.[6] Theoretical terms — such as mass, electron, force — are neither empty metaphysical results of the linguistic formulation of scientific theories, nor can their reference be "fixed" once and for all by so-called bridge relations or c-rules that are supposed to give them "observational" content. Rather, the meaning and reference of these terms slowly emerge during the various interpretative stages of science via the interpretative and empirical models of the theories containing them. (See Chapter 3.) Craig Dilworth expresses related ideas. He (Dilworth, 1994, p.155) writes:

> In seeking to understand why the laws of science take the form that they do, the scientist attempts to conceive what the reality underlying them must be like. He [sic] thus constructs an idealised model, which has as its *source* that which he [sic] feels he [sic] does understand ... The model should be constructed so as to depict a physically possible, albeit idealised, reality, whose existence would naturally manifest itself in the laws requiring explanation. ... a scientific model represents an ontology the nature of which may be taken as being responsible for the epistemology we associate with scientific laws. And ... the very fact that an explanation of such laws should be tentative necessarily suggests that that aspect of reality which is responsible for them is not open to direct inspection. Thus, if we consider models which depict these (perhaps temporarily) hidden aspects of reality as constructing the essence of scientific theories, we can characterise theoretical terms as terms used in referring to those entities in the real world (should they exist) as they are depicted in such models.

Now, continuing with Hacking's list, we have to ask whether science is cumulative or not. From my comments on this issue in Chapter 2 at least the following seems to be implied. Science is cumulative, yes, but in a certain specialised way. The metaphor of a clock[7] nicely illustrates this "special" kind of accumulation. Take the hour-hand to represent scientific theories, the minute-hand to represent the various context-specific models, and the second-hand to represent empirical data at the level of reality. I claim that the process of science is similar to the speed of the hands of the clock: theories change very slowly, models more quickly, and empirical data (observations etc.) the quickest, while definitive relations hold between these three aspects of science. (Maybe shifting to a new disciplinary matrix is analogous to buying a new clock.)

Each model of a given scientific theory is subjected to extremely strict testing against reality — by whatever form of testing is applicable to the model in question —

and it is seldom that a model withstands these tests without at least being modified in some way, i.e. without the intention shifting to some other model(s) (of the same theory) with better "empirical fit". Theory changes usually occur only when the possibility of changing and modifying the relevant models of the theory concerned has been exhausted. But even if a specific model of a theory is "discarded" the possibility of transforming that model and using it for other purposes in another interpretation of the same (or another) theory always exists. (Think of epicycles — the precursor of Fourier analysis — and the ether — the precursor of fields in space-time.) Thus the different levels of the scientific process are inherently related and it is this fact that guarantees the continuity of scientific knowledge.[8]

So far so good. How about Hacking's point on the deductive structure of science? Well, this is a difficult one. In my terms, a theory consists of a deductively closed set of sentences (in some appropriate language)[9], i.e. the theory contains all the sentences deducible from the axioms of the theory. In model-theoretic terms, the generalising activities of abstraction that take place at the beginning stages of the formulation process of a scientific theory are in direct contrast to the particularising activities of interpretation during the applicatory stages of science.

Next, we look at the nature of scientific terminology. Scientific terminology obviously tries to be as precise as possible, simply in the sense that it would seem a bit funny to be using vague ambiguous terms to formulate so many hours of detailed exacting research. But, if by "precise" is meant something like "having only one obvious absolute meaning or interpretation", I cannot agree, because such an approach totally ignores the model-theoretic implications of the scientific theory as a generalising statement. I am referring here to the various interpretations possible of one scientific theory by virtue of its abstract nature. The meaning and use of scientific terminology are also linked to specific models and are thus perhaps more contingent than some philosophers would like, since theoretical terminology has model-specific meaning and reference.

Finally let us consider the last point on Hacking's list, namely the contexts of justification and discovery. I agree with Kuhn that these contexts cannot really be separated. Even if they are taken in the traditional sense of the context of justification being about and using so-called "internal" features of theories, and the context of discovery being about and using factors "external" to scientific theories[10], the constant interaction and overlap between these contexts during the process of science become obvious in model-theoretic terms. It is possible to see that justification will perhaps play a larger role in the movement from the theory via its various models to systems in reality than in the initial development from an aspect of reality to the theory via the intended model. And, surely one could say that discovery will be very important in the initial stages of theory formulation, but so will it be in the final examining of systems of reality giving reference to the terms in the model(s) of some theory. Again, it is the interconnectedness of these stages or levels that makes it very difficult to separate these contexts.

It is however undeniable that both have important roles to fulfil in the process of science as a whole, as is specifically illustrated when the question of the truth of a theory in some interpretative model is examined. Scientists work within the context of the logic of justification when they construct the conditions and structure of the interpretation they work from in such a way that the theory (being the deductively justified consequences of axioms) has the best possible chance of being true under that interpretation. However, they work in the context of the logic of discovery when they determine the "truth" or "empirical adequacy" of the interpretation itself — that is the "fit" of the interpretation to some real system — because here they have to do with a multitude of different (often entertwined) tests (the results of which are represented in empirical models), varying from the empirical to the mathematical according to the specific interpretation they work from.

4. THE PROCESS OF SCIENCE: PARADIGMS AND MODELS

Thomas Kuhn's *The structure of scientific revolutions* (1970) represents a watershed in philosophy of science in terms of views concerning the nature of the scientific process, culminating in claims about the nature of science's progress, that cannot be ignored in a realist analysis of science such as the one offered here. In what follows I shall briefly discuss the nature of scientific progress in terms of a model-theoretic realist account of science, to show how Kuhn's revolutionary periods may be interpreted and accommodated. I shall also discuss the process of science in terms of the distinctions and common features between the Kuhnian notion of "paradigm" and my notion of "conceptual (interpretative) model". I shall mainly concentrate on the Kuhn of *The structure of scientific revolutions*, simply because that, in the end, remains the "bible" for Kuhnian philosophy of science.

Is the true nature of science revolutionary? Or is the process of science cumulative? Can there be revolutionary periods as well as cumulative periods in the development of science? Is the notion of paradigm essential to the idea of revolutionary change in science? Is the notion of model essential to the notion of cumulative scientific progress?

In *The structure of scientific revolutions* Kuhn (ibid., p.1) advocates a shift of focus in historical and philosophical studies of science from studying "finished scientific achievements" (ibid.) to concentrating on the "historical record of the research activity itself" (ibid.). He (ibid., p.2) is arguing against the conception that the content of science is somehow "uniquely exemplified" by observations, scientific theories and laws, and so against a rather naive interpretation of the notion of "development-by-accumulation" (ibid.). Believers in this view of scientific progress have the task of showing that once current views of nature, such as Aristotelian dynamics and phlogistic chemistry, are, as a whole, neither less scientific nor more the result of views peculiar to specific scientists, than later theories are.

Indeed one of the most powerful arguments against scientific realism — sometimes referred to as the so-called "argument from scientific revolutions" — centres around the seeming contradiction implied by the following question:

- How is it possible, if scientific realism is true, that the history of science abounds with examples of theories that have had great predictive success, but that were eventually "unmasked" by later science as being "unacceptable" or simply "false"?

There are of course various ways in which the argument from scientific revolutions can be approached. The one that seems to me to be the most plausible relates to discussions in recent philosophy of science by philosophers like Nancy Cartwright, and also the non-statement advocates such as Patrick Suppes, Bas van Fraassen, Ronald Giere, and a few others. In this context, I approach the problem of the nature of scientific progress by means of the conceptual system of models of theories that I have explained in the above. I claim that the role of so-called "scientific revolutions" in scientific progress may be interpreted as less interruptive and more continuous if the suppleness of this stage in the scientific process (i.e. the stage at which models are constructed) is allowed to influence our notions concerning scientific knowledge.

Now, Kuhn wishes to focus rather on the historical acceptance of scientific knowledge in a particular time than on any timeless contributions from an earlier science to the present one. To do this it seems to be important that any study of a specific scientific practice should be conducted from the point of view of the scientists working in that particular community at the time and not from the point of view of modern or current science. In this sense it seems likely that such a study may offer nothing more than explanations for the "internal coherence" of theories and for the reasons why they are deemed to offer the "closest possible fit to nature" at a specific time. Kuhn (ibid., p.4) stresses in this sense that it is indeed the "incommensurable ways of seeing the world and of practising science in it" (ibid.) that differentiate between various "schools of science" and not simply methodological failures or differences. He (ibid., p.118) claims that:

> At the very least, as a result of discovering oxygen, Lavoisier *saw nature differently*. And in the absence of some recourse to that *hypothetical fixed nature* that he 'saw differently', the principle of economy will urge us to say that after discovering oxygen he worked in a *different world*.

Kuhn goes on to claim (ibid., p.7) that "revolutions" do not simply imply the addition of one more "item to a scientist's world", but rather result in important changes in the scientific processional infrastructure. They cause the revision of the evaluative measures concerning experimental procedures that are in place at a given time in a given community, change the familiar ways in which entities were conceived or conceptualised up to that time, and ultimately, result in a shift of the "network of theory" through which the given community "dealt with the world" up to that point in time.

In what follows, I shall address the following questions related to the above:

- *Is* the content of science wholly, and uniquely — or even mainly — given by current observations and scientific theories and laws? In this context, what is the role of models in the process of science?

- If science is studied in the way advocated by Kuhn in *The structure ...*, how exactly should the nature and role of "paradigms" be understood? I.e. what is the difference — if any — between the role played by models in the process of

science and that played by paradigms?
- What is the meaning of Kuhn's "different worlds"?

In his best anti-realist manner, Kuhn (ibid., p.7) claims that

> Scientific fact and theory are not categorically separable, except perhaps within a single tradition of normal scientific practice. That is why the unexpected discovery is not simply factual in its import and why the scientist's world is qualitatively transformed as well as quantitatively enriched by fundamental novelties of either fact or theory.

I disagree with Kuhn about the reason why an unexpected discovery is, as he (ibid.) claims "not simply factual in its import". He offers the inseparability of scientific fact and theory as justification for this claim. I would, rather than entwining "real" facts and "scientific" theories from the start, instead speak of the layered nature of the process of science and consider each layer and its active factors in turn, and only then go on to study the connections and relations between layers, and between (some of) these layers and aspects of reality.

In this way, at least a conceptual distinction between science and reality becomes possible, and I see that as a plausible way in which to rescue some form of realism and also to escape the overwhelming (epistemological and ontological) relativism implied by the kind of constructivist approach Kuhn advocates. Think again of Bhaskar who states in *A realist theory of science* (1978) that the relationship between science and reality seems problematic only if one either accepts the social character of science, but denies that its object of study is independent of all social activity (the epistemic fallacy), or if one accepts the independence of reality, but denies the social nature of science (ontic fallacy). While Kuhn seems to be guilty of the first kind of fallacy, it is worthwhile to point out that the ontic fallacy (which seems somehow most compatible with the view concerning progress-by-accumulation) is *not*, however, the only other option open to us (as shown earlier in this chapter). Acknowledging that reality exists independently of the enterprise of science, need mean neither the denial of the social nature of science nor succumbing to views as narrow as those focussing only on the "natural", "fixed", or "objective" nature of the object of scientific study.

None of the above implies that for instance theories explaining and describing light as photons (quantum-mechanical entities that exhibit both characteristics of waves and of particles) are not scientifically "more advanced" or "better" than Huygens's theory of light, or theories based on Fresnel's and Young's theories that claimed light to be transverse wave motion. Neither does it imply that I somehow view Planck's, Einstein's, and Feynman's theories about light to be out of reach of further scientific activities or criticism. Rather, science progresses and scientific knowledge is cumulative in the following way.

As mentioned often, scientific progress is made at different speeds at the different levels of science. For instance, the *model* of Newton's theory of our solar system which works with seven planets can still be said to refer, model-theoretically, to an aspect of reality. The question of the truth of a theory is model-specific in the sense that it depends on the satisfaction of truth criteria which may differ from model to model and are satisfied differently in different models.[11] The general theory of relativity was formulated by Einstein (and Hilbert) in 1915. For more than 80 years now physicists have been constructing literally dozens of different types of models — all models of precisely the same theory — to fit the experimental and observational data about the space-time structure of the real universe, but also to fit individual or even cultural preferences, such as a passionate belief in, or an equally passionate aversion to, the idea of a Big Bang. Theory changes usually occur only when the possibility of changing and modifying the models of the theory concerned has been exhausted, which confirms the continuity of scientific knowledge. Think again of the rotation of the orbit of Mercury. Now, in this sense, I agree with Kuhn (as mentioned in Chapter 2) that neither the content of science nor any system in reality should be claimed to be "uniquely exemplified" by scientific theories from the viewpoint of studies of "finished scientific achievements".

Moreover, as mentioned before, any successful interpretation of a theory in a model is guided by contingent conditions which are context-specifically constructed. The range (or content) of these models thus cannot be established *a priori*. The formal (or *a priori*) property of scientific progress we as philosophers can say something about is the conceptual process of (re)formulating and applying theories in terms of (many different) models constructed to interpret the theory and having the ability to link systems in reality (via an embedded empirical model) to the theory itself. (This model-theoretic framework makes possible the verisimilitude enterprise of comparing theories as to their truthlikeness (of which more a bit later).) In this framework it is still possible to show that the progress of science is indeed accumulative.

The problem with Kuhnian revolutions is that they imply more than an epistemological shift, they require an ontological shift that necessitates discontinuity. Kuhn's "different" or "incommensurable" worlds address both the problematic issue of the invariance of (the order of) nature and that of the neutrality of science (or of the language of science). And that is at least one of the reasons why incommensurability has proved to be such an arresting notion — it concerns the status of both of the two poles of scientific realism.

Kuhn (ibid., p.111) states that "[e]xamining the record of past research from the vantage of contemporary historiography, the historian of science may be tempted to claim that when paradigms change, the world itself changes with them", although he qualified that later (in the Postscript) by saying that it is not "the world" that changes with a change of paradigm, but rather it is that scientists afterwards work in a "different world". He (ibid., p.121) wishes to argue against the Cartesian belief that "what changes with a paradigm is only the scientist's interpretation of observations that themselves are fixed once and for all by the nature of the environment and of the perceptual apparatus" (ibid.). In this sense of course, Priestley and Lavoisier both saw oxygen, and Aristotle and Galileo both saw pendulums, although in each instance, they differed in their interpretation of what they had seen.[12]

Now, since *The structure* ... Kuhn has often referred to the process by which later meanings are produced from earlier ones as a "process of language learning". By 1990 he had however found this metaphor too inclusive and had started concentrating[13] instead on the meanings of restricted classes of terms. These terms are taxonomic or kind terms, like "dog", "cow", "gold". They have two characteristic properties:

- They are identified as kind terms by virtue of lexical features such as taking the indefinite article.[14]

- They are subject to what Kuhn refers to as the "no-overlap" principle — no two terms of this kind may "overlap in their referents unless they are related as species to genus" (Kuhn in Tauber, 1997, p.233). For example, no dogs are also cows, a bangle is not a ring, and so on. Encountering a cow that is also a dog, would mean, in terms of this rule, that part of a taxonomy will have to be *redesigned*, and *not* that it is simply the set of category terms that has to be broadened or enriched.

To be able to describe reality at all, it seems obvious that in these terms, then, some kind of lexical taxonomy should already be in place. Moreover for communication to be possible, shared taxonomic lexicons have to be in place, because if different communities have taxonomies that differ in a certain local area, then situations may occur where statements made in the one will not be expressible in the other. Kuhn (ibid.) claims that the only way to bridge such a gap would be to breach the rule of no-overlap, and so incommensurability becomes a kind of untranslatability. It would be possible, though, for members of one community to learn the taxonomy employed by members of the other, but that would not mean that they would be able to translate terms from one taxonomy to the other, and, moreover the cost of bilingualism is that the particular community in which discourse is occurring at the time, has to be kept in mind throughout, to escape re-instating the original threat to communication. In the scientific context, Kuhn (in Tauber, 1997) makes it clear that periods of revolution will, in terms of the above, then be episodes which require local taxonomic change. All this is to say that there are times in the development of science when fundamental change in taxonomic categories is involved (for example, think of the Copernican and Ptolemaic taxonomies).

The interesting thing, in this context, about a model-theoretic interpretation of the process of science is that it can support certain changes of this kind, without subscribing to a discontinuous view of scientific progress. It seems obvious that there are "degrees" of incommensurability in the above sense — surely the "gap" between the Aristotelian view of science and the Einsteinian one is bigger than that between the Newtonian and the Einsteinian. In other words, it seems as if the nature of episodes of progress depends very much on the specific paradigms or disciplinary matrices involved. And, it is this aspect of the scientific process that may become a lot clearer if the mediation of models between theories and aspects of reality as a kind of magnification of scientific change is acknowledged, because the "decelerating" influence models have on this entire process helps to highlight the different factors of change and their different nuances in different situations.

Kuhn's emphasis on the directing role that paradigms play in science's processes is thus affirmed in a model-theoretic account of science, although in such an account the (extremely) slow pace of change of paradigms (in terms of disciplinary matrices) does not necessitate Kuhnian incommensurability. Given the overarching nature of disciplinary matrices, a change in disciplinary matrix might indeed be a revolutionary event, but the continuity models ascribe to science's processes is usually sufficient to render these events less interruptive than they are perhaps portrayed in Kuhnian terms.

The best way in which to interpret incommensurability in model-theoretic terms is perhaps then to acknowledge the — very gradual and slow — change of disciplinary matrix in the total process of science, but under the following conditions. The incommensurability should be understood to be about change in methodology, interpretation of observations, and general application of scientific knowledge. In other words, incommensurability (in a weak sense) may be more characteristic of model change than it is of change in disciplinary matrix. One may without real difficulty acknowledge that it would be difficult for Niels Bohr to work within the Einsteinian deterministic disciplinary matrix and get the same results as easily as he did, working in the context of the Copenhagen interpretation of quantum mechanics. However, Bohr and Einstein (Sachs, 1988) often had discussions about their different approaches. It might be that the more steadfast nature of disciplinary matrices that accounts for their slow change rate renders the conditions, detail of methodology, and implications of results of separate models less incommunicable as some believers in naive interpretations of incommensurability seem to claim.

A model-theoretic realism implies that a naive interpretation of incommensurability is a philosophical notion that has no real bearing on the process of science. Such a model-theoretic account of science makes it possible to examine technically various formal relations between different theories and their interpretations. Examples of such relations are:

- The interpretability of language L_1 in language L_2 — in which every primitive term of L_1 becomes a defined term in L_2.

- The interpretability of theory T_1 (in L_1) in theory T_2 (in L_2) — where L_1 is interpreted in L_2 and all the sentences of T_1 as interpreted in L_2 become sentences of T_2.

- Limiting interpretations — where terms of L_1 become, when some limit is taken, equivalent to terms of L_2. For instance Einsteinian mass (which depends on velocity) becomes Newtonian mass (which is independent of velocity) in the limit when the velocity is very small compared to the velocity of light.

Aspects like these and many more similar considerations must be taken into account when considering whether two theories are incommensurable or not. Only then could one eventually come to a definition (or probably different definitions) of what "incommensurability" may mean, and then continue to treat it technically.

Kuhn wants however to stress that, more than merely interpretational and methodological change, revolutions also result in change of disciplinary matrix, the theoretical and meta-theoretical background against which the scientists are working, and everything that goes with that. In model-theoretic terms, however, revolutions take place *only* in cases of disciplinary matrix change, and not in cases of model change.

The phenomenon of incommensurability should thus not necessarily be linked to revolutionary changes. It is possible to have two different models of the same theory within the same disciplinary matrix, while different meta-theoretical orientations do not prohibit scientists from understanding each other.

It seems — especially in later writings — as if Kuhn's objective with arguing about the variable sensory experience of scientists differently placed in history and differently placed with respect to alternative paradigms that form the "different worlds" in which they practice science, is still to somehow recover the "one world". These kinds of "different" worlds, however, have to do with *science* and its practice. It does not directly have anything to do with Nature, the "one" world, or however one decides to refer to it. It is, as Kuhn (1970, p.121) puts it, "the nature of the environment" and the "perceptual apparatus" that "fix" observations. It should be noted though that interpretations of observations can never be fixed — they can be refined, amended, sharpened, and changed in whatever way, at any minute in any number of ways for any number of reasons. (This is the important message of Section 2.7.) Secondly, the "nature of the environment" and the "perceptual apparatus" that shape interpretations of observations, are part of the disciplinary matrix within which the observations are made as well as the specific "view of the world" it offers.

Reality and our observations of certain of Nature's features are two different things. Nature and its features — such as gravitation — are independent of scientists and their actions, while observations of aspects of reality are not. As noted, it is important to distinguish between Nature and science; the one given, the other constructed. The always present danger of blurring the conceptual boundaries between Nature and science sometimes ends in a conflation of Nature and science, caused, it seems, by science's directedness towards reality. And this somehow implies, it seems — in a social constructivist kind of way — that science will have some kind of transforming influence on reality.

That is the problem: science may change our *conceptions* of reality (think of the role of background information in shaping these images). However, the features of Nature that we call gravitation have always been related to the features we call motion, position, mass, and acceleration and will always do so. The fact that Newton formulated a very workable, successful version of these relations does not change them in any way. Newton's formulation enabled — and still in limited domains enables — us to understand and employ the phenomenon of gravity and its manifestations, but our understanding will not change the working of these features — except in our models of it, in the "artificial environments" Cartwright likes to speak of.

This is no great disaster though, except if one takes this to mean that scientists have no knowledge about Nature. But it does not mean this, because "having knowledge of Nature" means exactly that abstractions are made from what is presented to our faculties of knowledge (however one wants to fill that in) and manipulating these abstractions conceptually and rationally. Science is the product of scientists' efforts to give a rational account of empirical observations and their implications.[15] Thus, the way some real system is will not itself be influenced by theory-change. Kuhn himself stated (in Tauber, 1997:243) that it is groups and their practices that constitute "worlds" (interpreted as the products of paradigms or disciplinary matrices) and that science is one of these "practices-in-the-world".

Therefore change in the disciplinary matrix from which scientists work may be triggered or influenced by theory-change, although this kind of change takes place even more slowly than theory-change itself. Think, for instance of Newton's method and how it differs from Cartesian mechanics, especially as far as the use of mathematics is concerned.[16] Descartes had tried to derive basic physical laws from metaphysical principles, while Newton insisted on basing his theorising on a careful examination of reality.[17] Finally, this kind of change may — and often does — lead to a change in the way in which scientists view reality, simply because different aspects of the system in reality they are studying are emphasised because of the change in purpose and method indicated by the matrix change.

The old debate between scientists aiming to "save the appearances" by "superimposing" mathematical relations on phenomena, and those who find explaining exactly why the phenomena are "there" more important, seems somehow to be lurking behind a lot of the issues raised in this text. Einstein spoke in this sense of "principled" and "constructive" theories, and considered his General Theory of Relativity as a principled theory in so far as it met the requirement of visualisability only after formulation. Quantum mechanics' development, however, was preceded by visual or pictorial interpretations, although the requirement of visualisability was rejected later on in formalising a number of basic quantum mechanical equations. Visualisation plays an interesting role in the search for atomic structure. The atomic model described by Rutherford in 1911 was closely analogous to the solar system. Some of Maxwell's theories of a more classical nature however gave rise to objections against this model. Bohr tried to eliminate these in 1913 by combining classical and non-classical approaches. Sommerfeld later elaborated Bohr's model and the analogy with the solar system was retained in the Bohr-Sommerfeld model.

But, by the 1920's Pauli and Heisenberg were rejecting this pictorial atomic structure. Pauli remarked that scientists should not try to shackle (the structure of) atoms with their preconceived opinions, but should rather adjust their concepts "in line with experience" (Sarlemijn & Sparnaay, 1989, p.7). In a famous paper, written in 1925, Heisenberg stressed that his basis for theoretical quantum mechanics is founded *only* on the relationships between quantities which are in principle observable. The Copenhagen interpretation of quantum mechanics supports the same approach: it is sufficient for the formalism adequately to establish the connection between the experimental results — no visualisation or physical interpretation of formal calculations are needed. A bit later, however, Heisenberg adopted a different attitude and began to search for physical interpretations that would "fit in" with matrix mechanics. This shows how the way chosen to solve a problem — which is in the first place determined by the relevant disciplinary matrix — may influence scientists' methods and goals — whether to strive towards "saving the appearances", or to have as objective to be able to show the connection between scientific conceptualisations and aspects of reality only after theory formulation has been initiated or even completed.

Scientific method should thus ideally provide a model-dependent model-modifiable strategy, because such a strategy (aided by the tools of non-monotonic logic in terms of minimal model semantics) offers — within a realist context — the possibility of modifying or amending our existing theories in the light of further research. The continuous nature of science is also confirmed, since the methodological principles of a strategy like this will themselves depend on the theoretical picture provided by currently accepted theories. Both our new theories and the methodology by which we develop and apply them depend upon previously acquired theoretical knowledge. And this fact about the cumulation of scientific knowledge — as well as science's various relations to reality — can (best) be supported and explained by a model-theoretic — realist — conception of scientific knowledge.

A last few remarks on the confirmation or succession of theories, focussing now, against the background of the above, on a model-theoretic vs a Cartwrightian interpretation of the role of models in the scientific process. Rueger and Sharp (1996, p.99) very adequately formulate the two basic choices about the nature of our theories with which Cartwright's (1983) model of explanation leaves us as follows: our theories are either

- simple and explanatory (in the sense of being classificatory with a wide scope), but not confirmable; or

- complicated and descriptive (phenomenological, with a narrow scope), but confirmable.

As I take these not as options for theories, but rather as characteristic of the different stages of the scientific process, I shall just point out that maybe some philosophers of science concentrate too exclusively on one aspect of the scientific process. Cartwright might be interpreted as feeling that philosophy of science is often too close to being exclusively philosophy of scientific theories to be able to offer a balanced view of the entire process of science (physics); and I agree, although I do not advocate getting rid of the role of theories as part of the referential process of science. Alan Musgrave (1981, p.381) claims that we

> ... do not falsify a theory containing a domain assumption by showing that this assumption is not true of some situations ..., we merely show that that assumption is not applicable to that situation in the first place.

In other words, finding a model in which a certain theory is false, does not necessarily imply the disconfirmation of the theory, at first it implies no more than that the theory is simply not applicable in that model.

Blaming the model rather than the theory, might sound indeed like using *ceteris paribus* clauses to protect theories against refutation. However, there is nothing *contrived* (or *ceteris paribus* for that matter) about this at all, it simply is the way science works (from a model-theoretic point of view, at least). For instance, the fact that it is extremely difficult to use Newton's laws to determine the route and behaviour of Neurath's bill blown about by the wind in St Stephen's Square (Cartwright, 1994b, pp.283-285) does not really say anything negative about Newton's laws, except that it is too difficult to apply them under certain circumstances.[18] The negative factor is rather that science then seems to have very little to say about wind-blown bills, which also is not really true, it is simply that this is a complex enough situation to warrant a new arrangement of abstractions, and so some kind of new formulation of a combination of certain fundamental laws, which might then lead to the construction of new models or amendments of old ones, which, in their turn, might turn out to be empirically adequate of the windswept hill, or whatever situation is being considered.

Finally, more often than not when the question of the continuity of science and the nature of the succession of theories in physics is addressed, the example of classical mechanics vs. quantum mechanics and relativistic mechanics is used. Let us briefly take a look, in Cartwright's terms, at the implications this example has for the role of models in the progress of science. Classical mechanics is usually viewed as a limiting case of relativistic and quantum mechanics. "Limiting case" here should be taken as meaning "an approximation to the new theory when its domain is limited to the domain where the older theory was (still is, for that matter) successful". Specifically, the equations of special relativity theory are taken to reduce to those of classical mechanics or non-relativist mechanics in the limit where the speed of light goes to infinity (which implies that there is very little that differs in the equations of these two theories if small velocities are being treated). And also then the equations of quantum mechanics are supposed to take on the form of those of classical mechanics in the limit where the quantum of action becomes negligible, as it is taken to do in the situation of macroscopic objects.

Now, although the notion of limiting cases is rather useful in the model-theoretic approach, even if perhaps mainly in practical terms of saving the time and effort that repeating the experimental activity involved in recreating the "old" theory's evidential base would cost scientists[19], the notion of approximation is too "naive" in a sense to really fit in easily enough (as mentioned before). The naivety of this notion lies in an illusion of *a priori*-ness it creates, in the sense of a definite controllable march towards final truth, which is just not part of a model-theoretic account of theories. The "hit and miss"-quality of the construction of models is then somehow lost[20], and that is one of the best sources of scientific creativity, I think, and I am in good positivist company too.

5. THE "ABSTRACT" AND THE "CONCRETE"

The special relationship between theories and models, and models and aspects of reality may be, finally, examined by considering Cartwright's characterisation of models in terms of fables (*Fables and models*, 1986). She starts the article with the following statement about her aims — the same as in *How the laws of physics lie* (1983), simply given from a different perspective —:

> I want to defend the view that the [phenomenological] laws may be true, literally true, yet they need not introduce new properties into nature. The properties they mention are already there; the new concepts just give a more abstract name to them ... we have no need to look for a single concrete way in which all the cases that fall under the same predicate resemble each other. What we need to understand, in order to understand the way scientific laws fit the world, is the relationship of the abstract to the concrete ... . (Cartwright, 1986, pp.56, 57)

The most important point about her discussion of the abstract and the concrete in this article (*Fables and models* (ibid.)) is the conclusion (already mentioned above) that laws or theories may somehow be said to be true (although she nowhere spells out that this actually means true-in-a-model), but they cannot be universal. She offers the finer detail of her arguments supporting this claim in *Nature's capacities and their measurement* (1989) by distinguishing between abstraction, idealisation, and concretisation, and offering a discussion of the problem of "material abstraction" (which refers to the problems arising from the fact that no universal recipe can be given for the movements from theory to concrete situations). Well, I agree that we *need* not believe that our scientific theories give us laws that are "underlying and determining what is going on" (ibid.). However, our bridges remain untouched by most storms, our aeroplanes do fly around the planet, our astronauts do walk on the moon, and so on and so on.

The interpretative development of scientific theories — from the theory to its interpretative models, and from there, possibly, via empirical models to some aspect of reality — as I see it, is a far more contingent issue than some philosophers would maybe care to admit — or be able to handle. I agree with Cartwright about speaking of truth only in terms of models — or the "artificial environments" we create for our theories' application — although I cannot see why that should prohibit me from keeping in the discussion something like Van Fraassen's (1980) notion of empirical adequacy to describe the nature of the "last (referential) jump" from interpretative model via empirical model to the aspect of reality concerned. Concentrating on — in my terms — the "jump" or movement from theory to interpretative model, Cartwright (ibid., p.58) quotes Gotthold Lessing's characterisation of fables as having the ability or potential to fill in the "graspable, intuitive content for abstract, symbolic statements".

Lessing (according to Cartwright (ibid., p.59)) also claimed that in order to make "a general symbolic conclusion" as clear as possible, it has to be reduced to the particular, so that it can be known "intuitively". That is what he means when he says the general can only become graphic or intuitive (*anschauend*) in the particular. (All very Aristotelian of course, with which I also have no quarrel.) Note (ibid.) that there is an ontological (the general *exists* only in the particular; where "exists" means "to be realised, and seen possibly to refer to some real system"), as well as an epistemological (the general becomes *graphic, anschauend* only in the particular; referring to conditions of truth — or empirical adequacy for that matter) claim being made here. Both these claims are addressed in my model-theoretic type of approach to science by the staggering of the various levels of scientific development, as well as by consideration of the nature of the relations and linkages between these levels.

Cartwright (ibid., pp.62 - 65) discusses Newton's second law of motion: $F = ma$, in these terms. This law is portrayed as an abstract "truth" — maybe theory would be a better choice of words — relative to claims about forces, motions, and masses. As it indeed is, and in the following way:

- (1) to be subject to a certain force, say F, is an abstract property;
- (2) $F = ma$ further claims that whatever object has the property in (1), also has

the property of having a mass and acceleration, the product of which gives the value F.

Scientists then have to figure out the nature of the typical situation in which real objects have these features. This implies that they have to figure out how to "fill in" the value(s) of each variable (see Cartwright (ibid., pp.62,63) for detailed discussion of this example) and then to construct a model in which each of the variables will behave according to their natures — e.g. "... we have the small mass m located at distance r from the larger mass M. Now we can look to see if the small mass moves with an acceleration GM/r^2 (since $GM/r^2 = F/m$). If it does, we have a model for Newton's law" (ibid., p.63).

She goes on (ibid., p.64) to set out a view regarding the abstract nature of theoretical concepts, which I think is close to what I accept as my view about the nature of these concepts (whether expressed in natural or mathematical language):

> On my account, force is to be regarded as an abstract concept. It exists only in the more specific forms to which it is led back via models ... It is not a new, separate property, different from any of the arrangements which exhibit it. In each case, being in this arrangement — e.g. being located a distance r from another massive body — is what it is to be subject to the appropriate force.

In other words, the abstract notions we are dealing with in scientific theories are not something over and above their concrete realisations — they do not in any way have "a life of their own" so to speak, floating around somewhere in some special kind of reality, they are merely part of the tools (Cartwright's point again in *The toolbox of science* (1995)) we use to formulate knowledge (or scientific) claims. The point is then simply that viewing the laws of physics as general (formal) claims and their concepts as abstract (and symbolic), comes down to acknowledging that Newton's $F = ma$ can be true of precisely those systems that it treats successfully. It however does not necessarily mean (Cartwright, 1986, p.67) that Newton has discovered a fundamental structure governing all of nature. Again, Cartwright's point about theories being *true* in limited domains, but *not universal* is illustrated.

Interestingly enough, this also shows the inaptness of a point Steve Clarke (1998) seems to want to emphasise; namely that there is some kind of tendency among realists (granted, he refers to fundamental realists, but my remark still is worthwhile, given the overall slant of his approach) to believe that one day, somehow, all fundamental laws will have been discovered and then the nagging open-endedness of the tiresome *ceteris paribus* clauses can forever be closed. The only kind of "open-endedness" I can see at issue in the process of science is the open-endedness of abstract linguistic expressions in the sense of the re-interpretability of theories as linguistic entities in various different interpretative models, and, moreover, it is as a result of suspending *ceteris paribus* clauses — if any are indeed present — that it is possible (within a model-theoretic scheme) to articulate links between fundamental laws and reality (or aspects of reality at least). The open-endedness that is referred to here can thus never be "closed" since that would mean turning to at least a non-statement approach as far as scientific theories are concerned, as well as considerably weakening the realist claims still possible in such a context.

In *Nature's capacities and their measurement* (1989), Cartwright works out the relations between abstractions, idealisations, and concretisations in far more detail, because, there, she discusses these notions in terms of causes and their complexities, and in terms of the capacities of nature. I shall briefly look now at how her notion of capacities clarify the tensions between the "abstract" and the "concrete" in science.

Cartwright amends her simulacrum account of explanation in *Nature's capacities and their measurement* (ibid.) by arguing that causal claims should play a central role in the explanations offered by science. This is an obvious continuation of her attack on the Humean characterisation of science (according to which explaining a phenomenon means showing it to be an instance of a general law or regularity), which she keeps focussed on arguments against the Humean attempt to reduce causal concepts to law-like ones (and finally to reduce these to regularities, or statements of association). She offers, instead, a metaphysics of enduring causal capacities[21]. She (1995c, p.292) claims that

> Laws in the conventional regularity sense ... must be constructed, and the knowledge that aids this construction is not itself again a report of some actual or possible regularities. It is rather knowledge about the capacities of [nature] and what these capacities can do if assembled and regulated in appropriate ways.

Dupré and Cartwright (1988, p.521) point out that no "right" sort of connections exist between capacities and properties of events. They (ibid.) write:

> Capacities are carried by properties. That is, you cannot have the capacity without having one of the right properties. But the same property can carry mixed capacities, and so the true complexity of the situation cannot be revealed by the association of properties.

And, moreover, since ... at any stage in [an] inquiry, there are always alternative sets of capacity that could account for the statistical data [under consideration]" (ibid., p.522), it is not possible — contrary to Hume — to find statistical data that can "settle" the truth of probable cases of regularity. In line with her empiricist's sympathies, Cartwright wishes to show that capacities can, however indeed, be measured.[22] She lengthily discusses the use of probabilities in this regard in quite a few of her more recent articles, as well as in *Nature's capacities and their measurement* (1989).

The primary problem that she wants to solve with her notion of capacities is whether — and to what extent — the laws of physics that are true in certain situations — that is in the "highly contrived environments of a laboratory or inside the housing of a modern technological device" (Cartwright, 1994b, p.281,282) — carry across to "systems, even systems of very much the same kind, in different and less regulated settings" (ibid.). She says

> The overall programme I want to urge is a careful and detailed philosophical story of the evidence about the boundaries of relevance ... for any of our ... fundamental laws. We have to allow for the possibility that they are true but not universal; exact but limited in range. (Cartwright, 1994c, p.293)

I too, have this aim. I want to realise it, however, by staying as far away as possible from any kind of metaphysical realism. Thus I cannot turn towards analysing the ontology of the things in reality that scientific theories may be "about" — whether referred to simply as the activities or behaviour of real phenomena, as the mechanisms or the tendencies of reality, the capacities of nature, or whatever else — but rather have to turn towards the ontology of *science* for my purposes.

The fact that capacities enable us to carry over information gathered in one set of circumstances to another, is the reason why capacity claims are not simply "higher levels of modality, but instead must be taken as *ascriptions of something real*" (Cartwright, 1989, p.158). I will briefly discuss what she has to say on this point, as it points to her realist tendencies with regard to capacities, while she remains at least uncertain with regard to the status of fundamental (abstract) laws. The problem is that if it is true that, as she claims now, fundamental laws are about the capacities of nature, as Chalmers (1993, p.201) points out, then they cannot describe sequences of events as well, and therefore they cannot anymore be taken to lie in the sense of *How the laws of physics lie* (1983).

Now, Cartwright (1989, p.178) argues that scientific methodology and its application presupposes that these capacities (or tendencies, as Mill calls them) are real. In other words, the only way in which fundamental laws can be taken to say something about reality, is if they are viewed as ascriptions of capacity. Although this doesn't change their *nature*, they still *lie* (according to her) however, because they are still Aristotelian abstractions. But, now, at least, they can be interpreted in some kind of realist terms. I fail to see the need for this. Whether the function of the fundamental laws of nature is to assign stable capacities to specific causes (Cartwright, 1989, p.179) or not, does not *really* decide whether these laws may have something to say about reality or not.

Alan Chalmers (1993, p.201) agrees with Cartwright that, if laws are supposed to describe capacities, then they cannot be taken to describe Humean sequences of events as well (Cartwright, 1989, pp.181ff.). But, then, fundamental laws do not really lie in the sense of *How the laws of physics lie*, because only laws taken in the sense of regularities could be said to lie in this sense (Chalmers, 1993, p.204). Which is close to what is worrying me, although I do not necessarily see scientific theories in terms of discoveries of regularities, because it is not <u>necessary</u> to think in that way within descriptions of the nature of theories and models and their interaction in my model-theoretic system. Moreover, we all know that science has never aimed at describing reality in all its complexity and fullness. The main problem in realist philosophy of science has always been, in a sense, to link the simplified versions of reality modelled by scientists to fit their theories (i.e. "scientific reality"), to the *reality* of reality as it were. Cartwright (and Roy Bhaskar) offer us ways in which to do this with their various descriptions of the "exportation of information" or bridging the "unsoundness argument".

David Spurrett (1999) writes: "Cartwright urges realism about capacities, and natures. Not being the kind of empiricist who would deny that, I am convinced by her arguments." I want to make it clear that I, too, see the point of at least some of her capacity-oriented arguments. My point here, for the purposes of this text, simply is that I do not want to be that kind of *realist*. I want to use the language of science — and find model theory the best available tool — to describe and explain why science refers to reality, rather than analyse reality's metaphysical components to address the same question. Spurrett (ibid.) continues:

> But she also wants to hold on to a quite empiricist notion of law as either regularity or probability [my point, earlier, that she wants to show capacities to be measurable], which means that in her terms we can only say that a law applies when we have constructed a 'nomological machine' (Cartwright, 1999, pp.49-74) which handles the ways different natures combine, when suitably isolated, to give rise to regularity and probability. ... [However] ... when one of Cartwright's nomological machines works it does so because there are true laws going into it in the first place, and ... some of these laws can be fundamental physical laws.

Model-theoretically, I agree, although perhaps for different reasons.

There is a fine difference in emphasis between Cartwright's account (of the process of science) and mine, but it is a very important one, because it relates to our attitudes towards the realism of our models. I would prefer to say, rather than simply claiming that theories cannot say anything about the aspect of reality that their models may be linked with, that theories can only explain in that little piece of reality that each of their models "refers" to (or rather might refer to). In other words, taking a previous example: the solar system model of Newtonian mechanics consisting of only seven planets, (used before Neptune was discovered) did indeed refer to the real situation. Although, in reality there were nine planets all along, the fact remains that it is quite possible to concentrate only on some of them and not on all at once. Whether and to what degree Newton's laws were empirically adequate when using this "restricted" model might seem to be a more difficult issue to deal with. But it is not really, since I claim — in agreement with Cartwright — that empirical adequacy (or truth) is a notion that can only be used meaningfully if linked with the model offering the relevant interpretation of the theory being considered at the time, and the relevant empirical model available (for use) at the time.

6. EMPIRICAL ADEQUACY

The main question that one tends to want to answer in a realist context is, of course, how exactly scientific theories "get to" reality. The problem is that there *is* no simple answer to that question. The conclusion drawn from claims like these should however not be that realism is untenable. A model-theoretic approach to realism proofs this, and more importantly, such a realism offers us the tools to examine — and make sense of — the *various* and *complex* empirical links between scientific theories and real systems. It may seem that — perhaps as a result of the many-to-many relations between theories and their interpretative and empirical models, as well as between these models and systems in reality — model-theoretically one merely ends up in the empirical reducts of some interpretative model of a given theory, rather than "in reality".

Well, what does it *mean* to be "in touch" with reality? I cannot see it meaning anything more than being in touch with the empirical practices of science. And, that is precisely what the empirical models embedded into a given theory's empirical reducts allow us to do. Of course, they allow this in a conceptual way, but then, the content of science — i.e. the set of its knowledge claims — is conceptual too is it not? A model-theoretic realism is a realism about objects in reality and the relations between them, although interpretative models (and the empirical models isomorphically embedded into them) are used to describe real systems.

The different accounts of science offered by the various non-statement approaches to the status of scientific theories all offer — among other things — variations on the theme of scientific realism. The model-theoretic tools these views are equipped with seem to offer a very good chance of, on the one hand showing that there are, indeed, certain empirical relations between the terms of theories and the entities in real systems, and on the other hand, defining the nature of these relations more and more precisely. (See Chapter 4.)

Sneed and his colleagues describe a class of (intended) applications of an empirical theory in terms of some class of partial potential models (i.e. theory-independent sets of observational reducts of potential models), and an empirical claim associated with the core of the theory in question. Such an empirical claim states that the class of partial potential models that satisfies the conditions set by the laws, constraints, and inter-theoretic links of the theory in question, is indeed in K, the theory's core.

Van Fraassen (1980, p.64) claims that

> To present a theory is to specify a family of structures, its models; and secondly, to specify certain parts of those models (the 'empirical substructures') as candidates for the direct representation of observable phenomena. The structures which can be described as experimental and measurement reports we call *appearances*: the theory is empirically adequate if it has some model such that all appearances are isomorphic to empirical substructures of that model.

In terms of Beth's state-space approach the link to reality is given via some satisfaction function between some (mathematical) state-space describing some physical system, and a set of elementary statements concerned with physical measurements. This means that the actual state of a physical system at a certain time may be given by defining some state-space representing the possible states of that system and some satisfaction function, which holds if the actual state of the system (described by some elementary statements) is an element of the domain of the relevant state-space.

Ronald Giere (1991, p.29) stresses that theoretical models themselves have no empirical content, but that theoretical hypotheses may identify elements of models with elements of real systems and then claim the relevant system exhibits the structure of the model. Evaluating the truth of hypotheses is a matter of statistical methodology (Giere, 1991), since he claims (Giere, 1985, p.80) a real system to "have the same structure as the model" if the system is similar to the model to specified degrees and in specified respects". These notions form the core of Giere's "constructive realism".

Wójcicki (in Humphreys, 1994) speaks of factual truth if the solution offered by some theory is true of the phenomena the theory wants to explain, as well as true in a model of the theory. This model should be a realisation of the theory and should represent the phenomena in question in all relevant respects. The following remark shows certain factors in Wójcicki's account of science, that a model-theoretic account also has. He (ibid., p.137) writes:

The fact that formation of a theoretical model presupposes formation of a model of the data as well as the fact that formation of a model of the data can be controlled by the requirement of consistency of the model with the corresponding theory are of key significance for proper understanding of the interplay between the data and the theories, and thus for proper accounting for both the corrigibility of the data and the falsifiability of the theoretical claims.

In explaining phenomena a model-theoretic realism tries to show how these phenomena, and our interactions with them, may be embedded — via some empirical model — into some empirical reduct of an interpretative model of some theory so that any real system exhibiting the phenomena in question may be within the reference of the theory's terms. Patrick Suppes's approach seems to me to hold the most promise as far as solving problems concerned with possible relations between theoretical entities, empirical data, and phenomena go. In my terms, proving the existence of relations of isomorphic embedding from empirical models into empirical reducts (of interpretative models) — which incorporates Suppes's hierarchy of models between, at the highest level, theories of experiments, and the notion of experimental design, closest to reality — offers a way in which to refer to the contingent and complex relations between real systems and mathematical models of theories in a precise way. These relations can otherwise only be examined and analysed *a posteriori* as part of historical studies of science.

From this viewpoint Cartwright's attacks ((1989), (1994a), (1994b), (1995a), (1999)) on "fundamentalism" seem rather empty. Very few philosophers of science still try to "close" the open-endedness[23] characteristic of models (and accompanying *ceteris paribus* clauses in Cartwright's terms). In an approach such as I am offering the abstract nature of theories, the "stabilising" and controlling nature of *ceteris paribus* clauses, the idealised nature of models, and the model-specific suspension of the said *ceteris paribus* clauses all fit into and can be accommodated in a logical reconstruction of the scientific process. Such a reconstruction can meaningfully address questions concerning the "truth" of theories, as well as their possible reference to reality. The use of minimal model semantics using total pre-orders to rank models in terms of default rules is an illuminating way in which to ponder the issues concerning idealisation versus concretisation, *ceteris paribus* clauses, over-determination of theories by their models (both interpretative and empirical), changes in exogenous and endogenous factors, and so on. (See Section 2.7.) After all, even Cartwright (1989) claims scientific theories to refer — albeit to what she terms causal capacities — if and only if the set of models in which the theory in question is true, is such that it has some further class of substructures where imputations of causal capacities are concerned, and is coextensive with the set of models actually used as working interpretations of the theory in question.

Returning to reflections concerning "touching reality", a clear manifestation of the empirical link between a model and some system in reality is given by the dimensional analysis in terms of the basic units in a derived physical quantity. According to the *Oxford concise science dictionary* (1996), a unit is the "specified measure of a physical quantity such as length [e.g. centimetre], mass [e.g. gram], time [e.g. seconds], etc., specified multiples of which are used to express magnitudes of that physical quantity" (ibid., p.751). The basic units of physical quantities are multiplied and divided to get derived units with dimensions, e.g. a unit of the form $L^pM^qT^r$, where L, M, and T indicate length, mass, and time, respectively, and p,q and r are (usually) integers. Examples: length (distance): $L = L^1M^0T^0$; mass: $M = L^0M^1T^0$; time: $T = L^0M^0T^1$; frequency, that is "per time": $1/T = L^0M^0T^{-1}$; speed: $L/T = L^1M^0T^{-1}$; acceleration: $(L/T)/T = L^1M^0T^{-2}$; momentum, that is mass × speed: $M(L/T) = L^1M^1T^{-1}$; force, that is mass × acceleration: $M(L/T^2) = L^1M^1T^{-2}$; energy, that is, work, that is momentum × speed = force × distance: $L^2M^1T^{-2}$; action, that is momentum × distance = energy × time: $L^2M^1T^{-1}$; power, that is force × distance ÷ time = work per time: $L^2M^1T^{-3}$.

The definitions (given in interpretative models of the measurement theory in question) of the basic units (and hence of the derived units) link these units empirically (calculations given by some empirical reduct of the interpretative model in question) to certain very definite aspects of reality. A second is the duration of 9 192 631 770 periods of a certain specific radiation emitted by a caesium-133 atom (that is the radiation corresponding to the transition between hyperfine levels of the ground state of this atom). A centimetre is the length of the path travelled by light in a vacuum during a time interval of $1/(2.99792458 \times 10^{10})$ second. A gram is one-thousandth of the mass of a certain platinum-iridium object kept by the International Bureau of Weights and Measures at Sèvres, near Paris in France.

In physics too we have extraordinarily accurate theories. Penrose (1997, p.51) reminds us that:

> In quantum field theory, which is the combination of quantum mechanics with Maxwell's electrodynamics and Einstein's Special Theory of relativity, there are effects which can be computed to be accurate to about one part in 10^{11}. Specifically, in a set of units known as 'Dirac units', the magnetic moment of the electron is predicted to be 1.001159652(46), compared with the experimentally determined value of 1.0011596521(93).

These instances also shows that indeed the highly regulated results of experimental situations may be (and <u>are</u>) "carried over" to the "complexities" of reality, quite successfully and without too much ado.

7. CONCLUSION: THE MEANING OF MODEL-THEORETIC REALISM

Adams (1959) is one of the first persons whom I know of who, in terms that may be interpreted in a realist way, described an empirical theory in terms of (i.e. as consisting of) two classes of structures: a class consisting of all the theory's "realisations", and a class consisting of all the intended applications of the theory in question. The latter class is merely a class of empirical structures (i.e. physical — or "real" — systems) of which the theory is (expected to be) true. As I see it, the problem is not only then to show that the theory is true of these empirical structures, but also to describe the relations — if any — between the "realisations" of the theory and these more physical structures making the theory true.

Now, as I have said often in the above, these are the two most difficult questions a realist, model-theoretically speaking, has to face. I propose answering these questions on the basis of — and analogous to, in a certain sense — the (formal) relations between a theory and its interpretative models (which I take to be Adams's "realisations"). In model-theoretic terms the answer to both the above "difficult" questions lies in the claim that a(n empirical) theory is "true" (i.e. "refers to" one) of Adams's empirical structures, because it is true (formally) in its interpretative model(s) within which we find the particular empirical structure(s) in question to be (isomorphically) embedded.

A model-theoretic approach to science supersedes and encompasses aspects of both the statement and non-statement accounts of scientific knowledge. Although in both the latter accounts — albeit in different ways — it seems that the notion of a scientific "theory" — however this notion is interpreted — may be given some realist interpretation at least, the unnatural rigidity of the statement approach's correspondence rules and the non-statement disconnectedness of scientific theories and their models do not allow for reference to reality in a satisfactory way. In a model-theoretic approach a scientific theory is a certain (deductively closed) set of sentences linguistically expressed.[24]

The conceptual embodiments of the contents of these theories are done via their models and the referential relations in question in a realist context are determined *both* by an empirical model, isomorphically embedded into some reduct of some interpretative model of the relevant theory, and so the nature of the real system in question; *and* by the nature of the interpretative model in which the theory is true. By maintaining such an encompassing interpretational link from the theories themselves all the way through their models to some real system(s) the complicated and changing character of science may be described and accounted for in a more adequate way than is perhaps the case with some statement and non-statement approaches to science.

A model-theoretic realism shows that over-determination — in a sense the converse of allowing only a "single causal story" (Cartwright, 1991b, p.386) — is a necessary characteristic of science, since the abstracting nature of the methodology of science specifically implies that other routes to the same conclusions are possible under a different abstraction from the same aspect of nature. As has already been pointed out in this chapter, the main assumption of such a realism concerning Nature is simply that it (i.e. Nature) exists independently of science. This basic condition is emphasised and worked out by the model-theoretic insistence on the roles that both science — in the guise of a specific interpretative model of a given theory having isomorphically embedded into it a certain empirical model — and Nature — in the sense of the characteristics of some real system satisfying the empirical results embodied by the specific empirical model in question — play in the processes of science.

No other metaphysical characteristic of Nature somehow worked into the mechanics of science is necessary to make sense of a scientifically realist picture of Nature. Rather the definition of the methodology and strategies — and aim — of science, model-theoretically interpreted, already takes care of all of that. And, a scientist can "know" that she is working with the "same phenomenon", even if using "different" theories, simply because of the possibility of analyses that a model-theoretic realism offers of tracing and articulating the different empirical links between different empirical models of different interpretative models in (perhaps) different theories.

As far as applications of model-theoretic analyses of scientific theories go, the following. One promising research area in philosophy of science that successfully uses model-theory and its interpretation of science is for instance the work being done in the area of verisimilitude. A model-theoretic realist account of science offers the tools necessary for the technical study of the "truthlikeness" or "closeness to the truth" of scientific theories. The study of verisimilitude had its beginning with Karl Popper's *Conjectures and refutations* (1963). This book contained a proposal for a definition of "theory y is closer to the truth (has larger verisimilitude) than theory x". It is interesting to note that Popper's approach to theories was firmly in terms of the statement approach — i.e. the view of theories in terms of deductively closed sets of sentences, although the study of the verisimilitude of theories may better be done in a model-theoretic context that has decidedly non-statement characteristics too. In 1974 David Miller and Pavel Tichý showed (independently of each other) that Popper's definition of verisimilitude is not usable, since according to it no two theories containing false sentences are comparable. Development after that has mainly been in the "non-statement" approach[25] — or rather in the spirit of my model-theoretic approach in which *both* models and theories play their equally essential roles. This work is exemplified by a large number of publications, of which the following is a selection of the more important and representative recent papers: Brink, C. & J. Heidema (1987), Burger, I.C. & J. Heidema (1994), Kieseppä, I. (1996), Kuipers, T.A.F. (1987), (1992), and (1997), Niiniluoto, I. ((1987), (1999)), and (1998), Oddie, G. (1986), Ryan, M. & P.Y. Schobbens, (1995), Zwart, S.D. (1998) and Zamora Bonilla, J.P. (1996). Due to limitations of space it is impossible to look into the relations between model-theoretic realism and truthlikeness here.

Secondly, the empirical aspect of science (to which, in the context of this text, only Suppes really paid adequate attention) is rather under-represented in philosophy of science, although more philosophers are currently writing on so-called "experimental realism" and using, for example in the case of Deborah Mayo, statistical methodology to address issues concerning theoretical parameters and hypotheses by means of experimental parameters and hypotheses that are identical to, or at least good approximations of, the theoretical ones.[26] A model-theoretic realist account of science, especially in terms of its views on over-determination, also offers a challenge to philosophers to look at this aspect of science with new eyes. Actually I think that current work done in non-classical logics, especially perhaps work done in non-monotonic logics and minimal model semantics, might serve to clarify much more than just the over-determination of theories by their various models.

A last few remarks now on the scope of model-theoretic realism. In Chapters 1 and 2 I have mentioned that the kind of model-theoretic account of science proposed here is probably applicable to the social sciences as well. It should be obvious though that different levels of both the formulative and the interpretative chains of such a model of science will be emphasised differently. For instance, the typical sociology of science fixation with analysing the nature of the particular scientific community in question, and the typical constructivist dependence on the theoretical framework in question are both catered for in model-theoretic realism. It might be that the construction of both the conceptual intended and the interpretative models of some theory would be a far more complex enterprise in the social sciences than in the empirical sciences. However, at the same time, a model-theoretic account of science is the one realist account available I know of that acknowledges and addresses the role of typical socio-constructivist features of the growth of scientific knowledge. Kuhn (1977, p.295) points out that if the term "paradigm" is to be fully understood, "scientific communities must first be recognised as having an independent existence". The same goes for understanding the development of scientific theories — from their origin to their applications — in a model-theoretic context.

Fine (1986a, p.150) claims that the process of science has a teleological side in the sense that

> ... the significance that realism attaches to science lends itself to the view that what science does ... is exactly what it aims to do. Thus we get the realist slogan that science aims at the truth, with the realist connection between truth and the *World* being understood.

In model-theoretic terms the aim of science simply is to offer certain idealised "insights" into the complex workings of "Nature". No statements about "absolute truth" or the unqualified "truth of scientific theories" are offered in such an approach and it is shown that such notions are empty and meaningless. Rather, systems in reality may be explained in terms of certain models interpreting a certain scientific theory. Scientific theories cannot be universally true, but merely true in (a) particular interpretative model(s) of it.

The slogan of a model-theoretic realism is "truth without universality". This is meant in the sense that it is the specific model-theoretic kind of truth that is at issue, and that theories are never examined for their relevance to reality in their stark linguistic terms, but always in terms of their (conceptual) interpretations in their various models. Theories in this sense are not viewed merely as general knowledge propositions, but rather as the means of organising systems of their models in such a way that certain systems in reality can be (empirically) "embedded" into these models.

In conclusion, the Aristotelian orientation of a model-theoretic realism can finally be given its rightful recognition. According to a model-theoretic realism such as mine, the general exists only *in* the particular. The meaning of the general and the particular are inseparably entwined. My criticism against the non-statement approach is based on this notion of the nature of general terms, since merely "giving" the theory "in terms of" its mathematical structures leaves out any real interpretation of the nature and role of general terms in the interpretation of science, and so, in my terms, actually renders the particular empirical interpretations of these structures meaningless.

In this sense, I prefer not to speak of the "universal" terms of a theory since these terms never have absolute interpretations. The terms of a theory are "general" in the sense that they are the result of certain abstractive manipulations of the object of scientific investigation at a certain time. Their (particular) meaning can be "given back" only by interpreting them in the limited context of the various interpretative models of their theory and, finally, by finding an isomorphic relation from some empirical model into some empirical reduct of the interpretative model in question. In this sense the notion of scientific "truth" becomes inextricably linked with that of reference, as it —

given its model-dependent nature — should be.

The essential link between the general terms of scientific theories and their interpretation in the various models of the theory regulates the rest of the referential process. While only the terms of the language which are "empirical" or "observational" in a particular empirical situation (observation or experiment) model-theoretically have qualified reference to reality, it is still the case though that the theory links its (theoretical) terms to observational terms — *conceptually* and *logically*. (See Chapter 2.) And, it is the linguistic formulation of the theory that is essential in the first place to make these links explicit and manageable.

Finally, let us consider the positioning of model-theoretic realism against at least some of the various forms of realism of contemporary philosophy of science, also considering the tension between empiricism and realism in philosophy of science. I shall refer here often to Van Fraassen's chapter entitled "The world of empiricism" in Hilgevoord's (1994) book entitled *Physics and our view of the world*, since the issues which I wish to discuss here seem to fit very well with the way in which he set out the theme of this chapter. Another reason is that Van Fraassen's views on the nature of the notion of empirical adequacy play an influential role in my portrayal of the relations both between empirical models and empirical reducts, and empirical reducts and interpretative models, and consequently the differences between the basic philosophical attitudes of model-theoretic realism and constructive empiricism are noteworthy.

The motivations of hard-core realists seem under certain interpretations of science rather uninformed and sometimes down right naive. Why is this so? Well, it seems that defenders of "strict" scientific realism of the metaphysical kind, aspire to answer questions that Van Fraassen (in Hilgevoord, 1994, p.114) refers to as "Why" questions. Van Fraassen (ibid.) claims that realists

> ... think of philosophy and science as jointly trying to uncover *what* is really going on in nature .... And at the same time, the realist sees science as aiming at real understanding of how nature works, and *why* it is the way it is. (My italics.)

Now, whether it is true that so-called metaphysical realists identify the former ("what") considerations with the latter ("why") considerations is not really what interests me here. The fact is that to me defenders of a realist philosophy of science should at most be concerned with "what" questions, and then from a second order point of view. That they should be (or are) interested in asking what goes on in nature, implies to me that they are interested to know first that science refers to *something*, and second (but not even necessarily), to what it is that science refers. This is not the same (at least not necessarily) as considering "why" questions, since philosophers interested in "why" questions first seem to want to explain nature or reality, before considering questions concerning reference. Apart from the fact that some members of this "why" category of philosophers have inherently strongly empiricist views with at least shades of instrumentalism, my view is that explaining nature is also first and foremost the worry of scientists, much more than it is that of philosophers of science. Worse, though, is the third possibility, i.e. that of realists considering answering the latter "why" questions, and then having only metaphysics to turn to.

I agree therefore with Van Fraassen (ibid., p.115) that the metaphysical nature of what has somehow become the traditional view of scientific realism might lie in an over-emphasis on the answering of "why" questions, and seeking the answers to these questions from science's supposed ability to postulate the existence of "deep" facts about reality. One of the main goals of this text is precisely to show that one does not have to be metaphysical in this sense to be a realist. Being a realist, to me, should mean simply that one is concerned with showing some relation of reference between the language of science and the world about which science supposedly "speaks". And this is how semantic realism is an aspect of scientific realism.

This might be all well and acceptable, but from the previous chapter we also know that I set great store by the relations (of empirical adequacy no less than in Van Fraassen's meaning of the phrase) by which we find some empirical model embedded into a particular empirical reduct at a certain time. This implies that some form of empiricism is lurking behind my realist facade. Well, of course there is. Fact is that I cannot see how else we are to determine in the final instance whether we have a "solid" or "valid" reference to some objects or events in some real system if not by at least making use of the results of empirical verification.

But, you see, Van Fraassen (ibid.) for instance, writes that the empiricist "... comes across as being 'against theory', calling us back to experience". Well, that obviously cannot be the case in my depiction of science. Allowing "experience" into the realist context is the only way in which the referential chain that a model-theoretic realist seeks, can be validated. You see, to a model-theoretic realist the realist question is not one concerning the ontology of reality, or of somehow establishing a corresponding mirroring (one-to-one) relationship between the ontology of science and the ontology of reality. Rather, the realist question is simply about whether scientific knowledge can be shown to be "about" reality. And, since we express our knowledge linguistically, scientific realism is semantic realism, not metaphysical realism. If philosophers want to answer questions concerning the nature of reality, given the meta-level, second-order nature of philosophy, they cannot turn to (object level, first-order) science for these answers, and so they have only metaphysics to turn to. But, model-theoretic realists do not ultimately want to answer questions concerning the nature of reality. Apart from the basic realist assumption that reality exists independent of us model-theoretic realism focusses on semantic rather than ontological questions.

Van Fraassen (ibid.) offers as an illustration of extreme empiricism the belief that "... things are entirely what they appear to be — and behind them ... there is nothing" (in the words of Roquentin in Sartre's *Nausea*). This, in essence, is what a model-theoretic realist, ultimately, want to believe too. Tracing realist relations between the linguistic and non-linguistic structures of science surely does not imply that there is anything "behind" the phenomena — in an unknowable *Ding-an-sich* way — in a certain real system that we are focussing on at a given time, but merely that a certain set of scientific conceptualisations may be shown to be linked to this real system in a certain (semantic) way at this time. We have only one way in which to express our knowledge (or perhaps I should say one "best" way), and that is to give it some form of linguistic expression. So, in the light of this, the realist task is first to trace the semantic links between the language of science and that "about" which science is.

The question that now appears is whether, if we show that there are relations of reference between the terms in the formal language of science and objects or events in real systems, this implies that those objects or events actually exist? Yes, is the answer, and immediately some readers will put model-theoretic realists back firmly classified with the metaphysical realists. At least the last sections of Chapter 2 should prevent them from doing this, though. It remains, finally, the empirical verification of the content of that which is embedded into a particular empirical reduct that proves that there "really" is "something" like the "thing" we referred to in the formal language. Why? Well, because we can show in a formal semantic way using the tools of model theory that there are referential links between some of the terms in our language and the empirical elements of some empirical model. And, frankly, this is all we have. This kind of empirical verification which I discussed in Chapter 2 is the "closest" we can get to reality.

To try to make the model-theoretic realist view a bit clearer, let us consider the traditional two views of language and its terms: nominalism and realism. Typically nominalists deny the existence of universals. That is they do not support the existence of the referents of general terms such as "flower", or "blue". They claim the resemblances between particular entities are sufficient to justify application of the same general term to all of such entities, and thus deny the need to appeal to any other (universal) entity to classify individual things. Realists traditionally have protested that such an approach still implies at least a tacit implication of reliance on universals, since the act of classification of things common to a certain set of individual entities implies a resemblance in some general respect. In other words realism with regard to particulars rests on some kind of realism about at least some universals, namely those that are the reputed referents of *bona fide* natural kind terms. More recently, nominalism usually implies the refusal to acknowledge the existence of abstract entities in general, whether they be particular or universal.

Nominalists thus have a rather less rigid approach than the typical (naive) realist attitude to the problem of reference in so far as they claim that general terms — i.e. "universals" — may exist only for the purpose of language and need not "correspond" to "real" entities at all. It is in this latter sense that model-theoretic realism might be said to exhibit shades of nominalism, albeit a very specifically qualified nominalism. Although an advocate of model-theoretic realism will claim that general terms refer in certain ways to particular things that (may) exist, she will emphasise just as strongly the re-interpretable nature of language. Thus, model-theoretic realism does not demand rigid unique mappings between language and the world, but rather concentrate on the different ways in which language might be "about" aspects of the world. This is the sense that I mean a model-theoretic realism to be "nominalist", i.e. against strict rigid one-to-one realism.

In the twentieth century philosophers such as Russell and the early Wittgenstein continued in the more conservative realist tradition which seems to subscribe to the belief that there is only one language and only one reality and that these two should (can) be linked in one-to-one relations.[27] Quine and Kripke also seem to have been fighting mainly on this side of the battle. The notion of a so-called "rigid" designator for instance supposedly captures the idea that throughout *all* possible worlds there are words that refer to the *same* individual. It is necessary that certain things are named in certain ways (or by certain terms).[28] I think it probable that somehow underlying all such approaches is a belief in some notion of absolute truth such that it is not allowed to arbitrarily ascribe "reality" to general terms. Nancy Cartwright's work on the lying laws of physics offers an interesting — albeit perhaps not a "typical" — case in point. The need to claim that these laws "lie" seems to me to be the direct result of some kind of notion concerning absolute truth. Why else should it *matter* that these laws can be concretised in so many different ways?[29]

During the last three or four decades there have been all kinds of reaction against the typical one-world-one-language account of the nature of abstract entities. One reaction is offered to us by the so-called "postmodern" notions of language and world that, in their extreme forms, seem to imply that Rorty's smashing of the "mirror of nature" eliminated all the chances — if there ever were any — to be able to describe reality — as "Nature" — at all. A more moderate reaction to the strict realist account of the reference of general terms is offered by a model-theoretic realism. The defenders of such a realism try to get some kind of a grip on the many-to-many relations between language and the models in which the sentences of the relevant language are true, and show that "truth" should be interpreted referentially in model-theoretic terms and not in terms of an infinite cancelling or postponing of meaning.

Within such an approach it is neither assumed — as in the case of defenders of formalism perhaps — that language expressions are merely "black marks" on white paper or air vibrations that can mean absolutely anything, nor is it claimed that abstract terms have any pre-determined reference or meaning. A model-theoretic realism escapes the extreme nominalist features of an approach such as that offered by so-called formalists because in model-theoretic terms not merely syntactical factors[30] are at issue, but also — most importantly — issues of semantics come into play. Tarski, Montague, and Adams may be viewed as some of the pioneers of this stream in philosophical thought with regard to the nature of abstract terms. There is however among model-theoretic advocates also some sympathy for Platonism. Perhaps Gödel offers the best example of Platonism in the philosophy of mathematics. In these — more Platonic — circles the mathematical tools most used are those of set theory and number theory.[31] Model-theoretically oriented mathematicians however give more attention to abstract algebras[32] — concerned with notions such as rings, groups, and vector spaces. These notions allow one to work with terms that do not necessarily always refer to the same things.

This is the basis of model-theoretic realism: the same language terms can refer to more than one entity in some models of the language (theory), and also the same object — or range of objects — in some real system can be referred to by more than one model, and, most importantly, these relations of reference can be traced, or articulated by using model-theoretic tools. Certain comments made by Jaakko Hintikka (1989) in a chapter entitled "Exploring possible worlds" in *Possible worlds in humanities, arts and sciences: Proceedings of Nobel Symposium 65*, may be used here to help fortify some of my main points.[33] He (1989, p.53) characterises the traditional nominalist-realist divide in terms of language as *calculus* and language as (the) *universal medium*. He (ibid.) explains the latter notion as essentially a denial of the possibility of "escaping" our language and looking at it and its logic from "the outside" (ibid.). This renders a model-theoretic realism "impossible" (ibid.) since the semantics — because of its supposed absolute character — of the relevant language becomes "inexpressible". From the more nominalist, "language as calculus", point of view however, the semantics of the relevant language becomes expressible to such a high degree that the notion of its interpretation as absolute becomes completely untenable.

Hintikka (1989, p.54) writes:

> The term 'language as *calculus*' is *not* calculated to indicate that on this view language would be a meaningless *jeu de caractères* — that is not the idea at all. Rather the operative word highlights the thesis that language is freely *re*-interpretable, like a calculus.

This, essentially, is what the notion of "epistemological relativism"[34] in model-theoretic realism is meant to imply too.

Hintikka (ibid., p.55) continues to point out that the re-interpretability of our language, according to the "language as calculus" view, implies that we can "... chose freely also the 'universe of discourse' as it [the relevant language terms] is designed to apply to" (ibid.). This again is much in accordance with my preferential (in terms of minimal model semantics) solution (Chapter 2) to "empirical proliferation". Given the abstractive selective nature of science — as discussed above in Chapter 2 and to be touched on again in Chapter 5 — and given this "free" choice of domain of interpretation[35], it seems obvious that a preferential model-theoretic realism of science is, at least in these terms, rather interesting.

However, again, as Hintikka points out too, this "choice" is not free to any absurd extent. I agree and my claim is that the "restraining" factor lies in the fact that a theory's terms — or some of them at least — may be shown to refer to some entities and objects of some real system *not only* because of the nature of the specific model interpreting the relevant theoretical terms at a certain time. Such an interpretation offers the necessary basic platform from which theoretical reference can be determined, but is by no means the final word. The final word is given by empirical data which, in the form of empirical models, make possible some relations of isomorphism from themselves into some empirical reduct(s) of the relevant model(s) on the one hand; and which, on the other hand, represent conceptualisations of features manifested by real interactions with the relevant real system.

Thus, it is in a sense both the "way" in which the world "is" (conceptualised partially via the specific empirical model in question) and the terms of the theory (interpreted via the specific interpretative model in question) that determine any articulation of possible reference. A model-theoretic realism thus displays a referential rather than a correspondence attitude towards the real systems science is "about". And, since the basic assumption of a model-theoretic realism is that science is about (human-independent) "Nature" via the descriptions and explanations of certain real systems offered by science's idealised models of the theory in question, any accusations of "rigging" the models of a theory such that the theory "really" refers — or anything akin to such claims, such as implied by supporters of Putnam's so-called "model-theoretic paradox" — are rather absurd.

Now, briefly returning to Van Fraassen and the empiricist-realist issue, Van Fraassen (in Hilgevoord (1994, p.124) writes:

> There are no necessary connections in nature, no laws of nature, no real natural bounds on possibility. Those ideas all resulted when philosophers projected familiar models onto the natural world. Really, nothing is necessary, and everything is possible. I mean this. ... What I reject is those philosophical ideas about where to turn for comfort. I am referring here to the realists' identification of understanding with knowledge of 'deep' facts about a reality behind the scenes of phenomena. Science is our paradigm enterprise of empirical inquiry and I value it very highly — but not as the acquisition of *such* knowledge.

Again, I agree. I still want to be a realist, though.

Van Fraassen (ibid., p.131) claims that realists focus too much on content and not on function and therefore they miss-represent the enterprise of science, and writes (ibid., p.132): "The primary commitment [in science] is to [a] ... method with ... [an] ideal of constant revolution and self-critique; all commitment to content is secondary", and adds (ibid., pp.132, 133) that

> If this conclusion about the primacy of method *vis-à-vis* content in science is correct, then realism has throughout mis-focussed the debate. For if realist metaphysicians reify content, then they do for science what the superstitious do for religion: *they avert attention from its significance to the vehicle of that significance*. What this means is that acceptance of science, and appreciation of its worth, does not require us to believe that it is true. On the contrary, the important point about scientific activity is not that it provides theories which every generation in turn can take as truth, but rather that it accustoms us to giving up our beliefs, changing and altering them, valuing them without being in bondage to them.

My comments on the nominalist aspects of model-theoretic realism gel quite interestingly with this view of Van Fraassen's, which, in essence, is of course an empiricist view. A model-theoretic realist interpretation of science calls itself "realist", because it wishes to show that science is indeed "about" reality. Its focus however is on tracing relations of reference, rather than on tracking (scientific) "truth", and in this its nature is functional or methodological rather than content-oriented. Moreover, the final stages of determining such relations of reference are still empirical or empiricist, since the empirical models of theories are taken as conceptualisations of experimental activities. However, it is not, as the positivists seem to have claimed, a theory of experiential or empirical elements that bridges the gap between descriptions of matters of fact and the general principles (theories) of science, but rather complex (semantic) relations of satisfaction.

Notice that reflections on whether these relations determine a "correct" or "true" representation of reality remain silly. The evolutionary nature (rather than "revolutionary" given the various stages of scientific processes and progress or development as portrayed in the previous chapter) of science, combined with the "contructedness" of science, are fundamental to a model-theoretic depiction of science. "Truth" is relative to specific models. Questions of truth can indeed only be settled by focussing on conditions of verification, but in the semantic sense of defining interpretations of the scientific language on specified domains of discourse.

So, there certainly are elements of conventionalism in a model-theoretic approach to theories, in the sense that "truth" is something that we "create" by our (pragmatic) choices (of interpretation), and not something dictated to us by nature. At the same time though, we understand that the "truth" of a theory in one model means the same as "truth" of a(nother or the same) theory in another — in other words the notion of truth is transcendental, in the sense of being effable even though it can only be given content in particular (different) contexts. This conventionalism however does not collapse into the kind of reductionism against which Quine had it in "Two dogma's" (1953). The range of verifying conditions corresponding to each statement of a language are not determinable *a priori*, but rather *a posteriori*, in the sense that the choice of (verification) condition rests on the nature of the interpretational domain of denotation for a given language, which, in its turn, is determined by extra-logical and extra-scientific factors inherent to disciplinary matrices and goals of theory application (among others).

It must be emphasised though that the kind of relativism that appears to be present here is an epistemological rather than an ontological relativism (see earlier sections in this chapter). Within a model-theoretic framework epistemological issues are addressed by issues of semantic reference. And, it is taken as a given that the reality that is bound to science and its theories through the semantic relations of reference is a reality that exists independently of human (scientific) constructs, i.e. there is no ontological relativism here. It is just that different interpretations of scientific theories refer to different aspects of this reality at different times. I have already elaborated on these thoughts in Chapter 2.

So, yes, a model-theoretic realist analysis of science is first a semantic analysis of the language of science. In this is retained the positivist emphasis on the importance of the relations between languages and their logical structure. Rather than stating that the *meaning* of a sentence is given by the conditions for its verification, a model-theoretic realist would claim that the *reference* (or realist content) of a sentence is given by the conditions for its verification.

A model-theoretic realism thus offers a conventionalist, nominalist, empiricist (in the broad sense of relating knowledge claims in some way to experience), constructed, referential depiction of science, and the "realism" at issue within this framework is far removed from naive "T"ruth-seeking one-to-one correspondence realism. When positivism was faced with the fact that no absolute verification of scientific theories through predictions that come out "true" is possible, it seemed that after all, there were only two ways out: to find some kind of epistemological (metaphysical) justification for scientific knowledge, or to characterise the meanings of as many parts of the language of science as possible. There is, however, a third way out: finding a method for interpreting the language of science without collapsing epistemology into either semantics or metaphysics. I claim that this is what a model-theoretic realism, together with the non-classical methods of non-monotonic logic in terms of minimal model semantics offer: an evolutionary realist epistemology of science semantically determined by tracing non-unique relations of reference between language and world.

In conclusion let us consider within a model-theoretic realist context the five theses of realism that Ilkka Niiniluoto (1999, pp.10ff.) formulates as part of his excellent discussion of the various forms of realism philosophy of science has to offer in his book entitled *Critical scientific realism* (ibid.).

- Ontological realism: "At least part of reality is ontologically independent of human minds" (ibid., p.10). Model-theoretic realism accepts this thesis.

- Semantic realism: "Truth is a semantical relation between language and reality. Its meaning is given by a modern (Tarskian) version of the correspondence theory, and its best indicator is given by semantic enquiry using the methods of science" (ibid.). Model-theoretic realism also accepts this thesis.

- Theoretical realism: "The concepts of truth and falsity are in principle applicable to all linguistic products of scientific enquiry, including observation reports, laws, and theories. In particular, claims about the existence of theoretical entities have a truth value" (ibid.). Model-theoretic realists accept this thesis, but only in terms of truth-in-models.

- Axiological realism: "Truth (together with some other epistemic utilities) is an essential aim of science" (ibid.). Model-theoretic realism has only one notion of truth, namely truth-in-a-model. That is why questions concerning truth are referentially interpreted and answered. Preferential model-theoretic truth is essential for scientific progress, but "truth" in an absolute sense cannot be considered as an aim of science, since the notion itself is without sense in a model-theoretic framework.

- Critical realism: "Truth is not easily accessible or recognisable, and even our best theories can fail to be true. Nevertheless, it is possible to approach truth, and to make rational assessments of such cognitive progress" (ibid.). Again, first, model-theoretic realism talks only of truth-in-a-model. The amendable or defeasible character of the "truth" of scientific theories is model-theoretically depicted not by "approaching truth", but rather by analyses of scientific progress in terms of minimal model semantics, especially in terms of the two default rules formulated in Section 2.7.

- Linked to the previous thesis, is the last one: "The best explanation for the practical success of science is the assumption that scientific theories in fact are approximately true or sufficiently close to the truth in relevant aspects. Hence, it is rational to believe that the self-corrective methods of science in the long run has been, and will be, progressive in the cognitive sense" (ibid.). Model-theoretic-realistically the self-corrective methods of science are also taken as evidence for past and future scientific progress. Again, though, truth is here interpreted in preferential model-theoretic terms.

Model-theoretic realism thus implies ontological and semantic realism, in order to be an epistemological (in terms of scientific knowledge claims) realism, showing traces of critical realism only in so far as the self-corrective methods of science are also accepted and taken to make rational assessments of scientific progress possible.

Again, this form of realism might seem weak by traditional standards. If, though, both truth and reference are interpreted contextually (i.e. model-specifically), <u>and</u> can still be shown to be intelligible notions that can be shown to hold without needing a traditional metaphysical analysis of the ontology of reality, we actually have a much richer form of realism here. Also model-theoretic realism is much closer to the actual nature of science than metaphysical realism, since no one-to-one relations between scientific language and the world are posited or needed for realising realist ideals. Tarski's interpretation of truth as reflecting a relationship between sentences in some language and interpretations of the appropriate language, rather than a property of a sentence is the best thing that happened to realism in the last 70 or so years.

NOTES: CHAPTER 5

[1] Some sections of this chapter are published as Ruttkamp (1999b).

[2] In a recent spirited defense of "causal realism" Christopher Norris (1997) also offers a criticism of the anti-realism implied by Van Fraassen's "constructive realism". (See Norris (1997, Chapters 6 & 7).)

[3] I have already briefly commented on this issue in Chapters 3 and 4.

[4] See Spurrett (1998) for a thorough discussion of this aspect of Bhaskar's transcendental realism.

[5] Speaking of the "contingent" nature of data is not meant to sound frivolous, but rather to refer to the verification of the validity of data via the various models of some scientific theory. (See Section 2.7 again in this context.)

[6] Here I agree with Kuhn's (1977, p.267) description of the "intimate and inevitable entanglement of scientific observation with scientific theory" that leads — as Kuhn (ibid.) also remarks — to a certain skepticism concerning the production of a "neutral observation language".

[7] I thank Professor Johannes Heidema of the Department of Mathematics, Applied Mathematics, and Astronomy at the University of South Africa for this metaphor.

[8] Nancy Nersessian (1984, p.153) writes that a discussion of a concept of meaning that would be adequate for scientific theories should be given in terms of the following two factors. The study of nature and an analysis of language, as well as an analysis of "actual scientific practices concerning meaning" (ibid.). She (ibid.) claims that it is impossible to separate questions of meaning from the "network of beliefs (theoretical, methodological, metaphysical, common sense) and problems (theoretical, experimental, and metaphysical) which provides the 'motive force' in meaning construction" (ibid.). She (ibid., p.156) continues to describe the meaning of a scientific concept as "a two-dimensional array which is constructed on the basis of its descriptive/explanatory function as it develops over time. I will call this array a 'meaning scheme'". (This reminds somewhat of the positivists's linguistic frameworks (see Chapter 3), and also of Davidson's (1984) "conceptual schemes".)

[9] See Chapter 2.

[10] See Carnap (1956b), and also Chapter 3.

[11] See Heidema, J. & H.J. Schutte (1978).

[12] In section X of *The structure* ..., Kuhn (1970, p.126) asks: "But is sensory experience fixed and neutral? Are theories simply man-made interpretations of given data?". And he answers: " ... Yes! In the absence of a developed alternative, I find it impossible to relinquish entirely that viewpoint. Yet it no longer functions effectively, and the attempts to make it do so through the introduction of a neutral language of observations [Quine] now seem to me hopeless". In the Postscript, Kuhn (1970, p.193) tries to solve his problem by drawing clear distinctions between sensory "stimuli" and "sensations" or "perceptions". He (ibid.) writes:

> "Notice now that two groups, the members of which have systematically different sensations on receipt of the same stimuli, do *in some sense* live in different worlds. We posit the existence of stimuli to explain our perceptions of the world, and we posit

their immutability to avoid both individual and social solipsism. About neither posit have I the slightest reservation. But our world is populated in the first instance not by stimuli but by the objects of our sensations, and these need not be the same, individual to individual or group to group".

And, he continues to say that it is because we have been conditioned to see a one-to-one mapping between stimuli and sensations, that we have such difficulty in recognising that the two viewers actually see different things. In reality, we should — and do — know that the same stimulus may produce very different sensations and that very different stimuli can produce the same sensations.

[13] See his article in Tauber (1997).

[14] Kuhn describes (Tauber, 1997, p.233) the meaning of these terms as "part of what one must have in the head to use the word properly".

[15] See also Cook, A. (1994). *The observational foundations of physics*. Cambridge University Press.

[16] Newton "did not produce mere mathematical constructs or abstractions that were devoid of any content of reality other than 'saving the phenomena', but he did create what he conceived to be purely mathematical counterparts of simplified and idealised physical situations that could later be brought into relation with the conditions of reality as revealed by experiment and observation" (Sarlemijn & Sparnaay, 1989, p.6). He also preferred synthetic geometry to Descartes's analytical geometry and even to his own calculus, because both the latter have levels of proof without any clear physical interpretation. Berkeley even referred to the infinitesimals in Newton's calculus as "the ghosts of departed quantities".

[17] Newton affirmed Aristotle's inductive-deductive method — he called it the "method of analysis and synthesis". Newton declared that "although the arguing from experiments and observations by induction be no demonstration of general conclusions, yet it is the best way of arguing which the nature of things admits of" (Newton, 1952, p.404).

[18] Note that, in principle, it *is* possible to deduce the movement of "Neurath's bill" *perfectly* from Newton's laws. The real system is in this case simply too complex to be able to actually make this deduction in practice.

[19] Cartwright refers to this too.

[20] As the criticism of "fundamentalist realist" theories of science aptly shows.

[21] This notion of capacities might remind of Popper's propensities. See Popper (1990).

[22] In addressing the testability of causal claims, Cartwright *uses* probabilities, while the Humean tradition *reduced* causal laws to probabilities. She says: "I defend a very different understanding of the concept of Natural Law in modern science from the 'Laws = universal regularities' account ... We aim in science, I urge, to discover the *natures of things*; we try to find out what powers or capacities they have and in what circumstances and in what ways these capacities can be harnessed to produce predictable behaviours. I call this the study of *natures* because I want to recall the Aristotelian idea that science aims to understand what things *are*, and a large part of understanding what they are is to understand what they *can do*, regularly and as a matter of course." (Cartwright, 1995c, p.277).

Ceteris paribus clauses can, however, it seems, not be escaped — "In order to generate a prediction [or, give an explanation] we must figure out how to combine the laws together and how to cash-out their *ceteris paribus* conditions — and we must do so in a way that takes into account the specific material circumstances of the situation under consideration" (Cartwright, 1995a, p.155). The way to do this then, is to assume the existence of capacities (as has already been pointed out) — "The point is that the fundamental facts about nature that ensure that regularities can obtain are not again themselves regularities. They are facts about what things can do" (ibid., p.156).

[23] See Clarke (1998) as well.

[24] Note that, in principle, the language in question need not be a formal language at all. Almost any kind of scientific linguistic expression may be formalised in some first-order language and its interpretations reconstructed in a model-theoretic way.

[25] Niiniluoto (1999, pp.139ff.) points out that for instance Ronald Giere's constructive realism in which a theory is true in a model, and the model is "similar" to a real system generates truthlikeness.

[26] Don't leave out Galison (1987) and (1997) who offers a treasure trove for the philosopher in this respect.

[27] Think of the notion of so-called "picture theories", and of logical atomism, for instance.

[28] This seems to be the basic idea of Kripke's *Naming and necessity* (1980).

[29] Of course she also is referring to the idealised nature of the models interpreting the fundamental laws of physics when she speaks of the falsity of these laws. That is however a different issue, addressed in Chapters 4 and 5.

[30] See Lewis (1970) for a good illustration of the syntactic scheme of things.

[31] Perhaps the reason for this is simply the historical psychological "feeling" that numbers and sets are somehow unique entities (floating around somewhere).

[32] Abraham Robinson (for instance 1986) developed the model-theoretic structure of algebra in this way.

[33] See also Hintikka (1997).

[34] See Chapter 5.

[35] Hintikka (1989, p.55) remarks that this domain can be so particular that it can be characterised as "... a 'small world', that is, a relatively short course of local events in some nook or corner of the actual world".

BIBLIOGRAPHICAL REFERENCES

Achinstein, P. (1968): *Concepts of science: A philosophical analysis*, John Hopkins Press, Baltimore.

Adams, E.W. (1959): The foundations of rigid body mechanics and the derivation of its laws from those of particle mechanics, in: Henkin, L., P. Suppes & A. Tarski. (eds.), *The axiomatic method*, North-Holland, Amsterdam, pp. 250-265.

Bachelard, G. (1934): *The new scientific spirit*, Beacon Press, Boston.

Backhouse, R. (1997): *Truth and progress in economic knowledge*, Edward Elgar, Cheltenham.

Balzer, W. (1982): A logical reconstruction of pure exchange economics, *Erkenntnis*, 17, pp. 23-46.

Balzer, W. (1985): The proper reconstruction of exchange economics, *Erkenntnis*, 23, pp. 185-200.

Balzer, W., C.U. Moulines, & J.D. Sneed. (1987): *An architectonic for science - The structuralist programme*, D. Reidel, Dordrecht.

Balzer. W. & B. Hamminga. (eds.), (1989): *Philosophy of economics*, Kluwer Academic Press, Dordrecht.

Beth, E.W. (1948/49): Analyse sémantique des théories physiques, *Synthese*, 7, 206-207.

Beth, E.W. (1949): Towards an up-to-date philosophy of the natural sciences, *Methodos*, 1, 178-185.

Beth, E. W. (1961): Semantics of physical theories, in: Freudenthal, H. (ed.), *The concept and the role of the model in mathematics and natural and social sciences*, D. Reidel & Co, Dordrecht, pp. 48-51.

Beth, E.W. (1963): Carnap's views on the advantages of constructed systems over natural languages in the philosophy of science, in: Schlipp, P.A. (ed.), *The philosophy of Rudolf Carnap*, Open Court, La Salle, pp. 469-502.

Bhaskar, R. (1978): *A realist theory of science*, Harvester, Sussex, 2nd ed.

Bhaskar, R. (1979): *The possibility of naturalism: a philosophical critique of the contemporary human sciences*, Harvester, Sussex.

Bhaskar, R. (1986): *Scientific realism and human emancipation*, Verso, London.

Bhaskar, R. (1989): *Reclaiming reality: a critical introduction to contemporary philosophy*, Verso, London.

Blaug, M. (1992): *The methodology of economics*, Cambridge University Press, Cambridge, 2nd ed.

Blaug, M. (1994): Why I am not a constructivist: Confessions of an unrepentant Popperian, in: Backhouse, R.E. (ed.), *New directions in economic methodology*, Routledge, London.

Boyd, R. (1994): What realism implies and what it does not, in: Worrall, J. (ed.), *The ontology of science*, Dartmouth, Aldershot, pp. 71-96.

Braithwaite, R.B. (1953): *Scientific explanation. A study of the function of theory, probability and law in science*, Cambridge University Press, Cambridge.

Brink, C. & J. Heidema. (1987): A verisimilar ordering of theories phrased in a propositional language, *The British journal for the philosophy of science*, 38, pp. 533-549.

Burger, I.C. & J. Heidema. (1994): Comparing theories by their positive and negative contents, *The British journal for the philosophy of science*, 45, pp. 605-630.

Carnap, R. (1936): Testability and meaning, *Philosophy of science*, 3, pp. 420-468.

Carnap, R. (1937): Testability and meaning, *Philosophy of science*, 4, pp. 1-40.

Carnap, R. (1947): *Meaning and necessity. A study in semantics and modal logic*, University of Chicago Press, Chicago.

Carnap, R. (1950): Empiricism, semantics and ontology, *Revue Internationale de Philosophie* (Brussels), 4ᵉ anneé, 11, pp. 20-40. (Reprinted in: Linsky, L. (ed.), (1952): *Semantics and the philosophy of language: A collection of readings*, University of Illinois Press, Urbana, pp. 208-228.)

Carnap, R. (1956): *Meaning and necessity*, Chicago University Press, Chicago, 2nd ed.

Carnap, R. (1958): Beobachtungssprache und theoretische Sprache, *Dialectica*, 12, pp. 236-248.

Carnap, R. (1966): *Philosophical foundations of physics: An introduction to the philosophy of science*, Basic Books, Inc., New York.

Carnap, R. (1975): Observation language and theoretical language, in: Hintikka, J. (ed.), *Rudolf Carnap, logical empiricist*, D. Reidel Publishing Company, Dordrecht, pp. 75-86.

Cartwright, N. (1979): Causal laws and effective strategies, *NOÛS*, 13, pp. 419-437.

Cartwright, N. (1983): *How the laws of physics lie*, Oxford University Press, Oxford.

Cartwright, N. (1986): Fables and models, in: Worrall, J. (ed.), *The ontology of science*, Dartmouth, Aldershot, pp. 191-204.

Cartwright, N. (1989): *Nature's capacities and their measurement*, Clarendon Press, Oxford.

Cartwright, N. (1991a): Can wholism reconcile the inaccuracy of theory with the accuracy of prediction? *Synthese*, 89, pp. 3-13.

Cartwright, N. (1991b): The reality of causes in a world of instrumental laws, in: Boyd, R., P. Casper & J.D. Trout. (eds.), *The philosophy of science*, MIT Press, Massachusetts, pp. 379-386.

Cartwright, N. (1993): Arsitotelian natures and the modern experimental method, in: Earman, J. (ed.), *Inference, explanation and other philosophical frustrations*, University of California Press, Berkeley.

Cartwright, N. (1994a): In defence of this worldly causality. Comments on Van Fraassen's *Laws and symmetry*, *Philosophy and Phenomenological Research*, 53(2), pp. 423-429.

Cartwright, N. (1994b): Fundamentalism and the patchwork of laws, *Proceedings of the Aristotelian Society*, 94, pp. 279-292.

Cartwright, N. (1994c): Is natural science natural enough? A reply to Philip Allport, *Synthese*, 94(2), pp. 291-301.

Cartwright, N. (1995a): Précis of "Nature's capacities and their measurement, *Philosophy and Phenomenological Research*, 55(1), pp. 153-156.

Cartwright, N. (1995b): False idealisation: A philosophical threat to scientific method, *Philosophical Studies*, 77(2-3), pp. 339-352.

Cartwright, N. (1995c): *Ceteris Paribus* laws and socio-economic machines, *The Monist*, 78(3), pp. 276-294.

Cartwright, N. (1997): Why physics?, in: Penrose, R. (ed.), *The large, the small, and the human mind*, Cambridge University Press, Cambridge, pp. 161-168.

Cartwright, N. (1998): Capacities, in: Davis, J.B., D.W. Hands & U. Mäki. (eds.), *The handbook of economic methodology*, Edward Elgar, Cheltenham, pp. 45-48.

Cartwright, N. (1999): *The dappled world: A study of the boundaries of science*, Cambridge University Press, Cambridge.

Cartwright, N. & J. Dupré. (1988): Probability and causality: Why Hume and indeterminism don't mix, *NOÛS*, 22, pp. 521-536.

Cartwright, N. & H. Mendell. (1984): What makes physics' objects abstract?, in: Cushing, J.T., C.F. Delaney, & G.M. Gutting. (eds.), *Science and reality. Recent work in philosophy of science*, University of Notre Dame Press, Indiana, pp. 134-152.

Cartwright, N., T. Shomar & M. Suárez. (1995): The tool-box of science, in: Herfel, W.E., W. Krajewski, I. Niiniluoto & R. Wójcicki. (eds.), *Theories and models in scientific processes. Poznán studies in the philosophy of sciences and the humanities*, 44, pp. 137-149, Rodopi, Amsterdam.

Chalmers, A. (1987): Bhaskar, Cartwright and realism in physics, *Methodology and science*, 20, pp. 77-96.

Chalmers, A. (1993): So the laws of physics needn't lie, *The Australasian journal of Philosophy*, 71(2), pp. 196-205.

Chang, C.C. & H.J. Keisler. (1990): *Model theory*, North Holland, Amsterdam, 3rd ed.

Chiang, A. (1974): *Fundamental methods of mathematical economics*, McGraw-Hill, New York.

Clark, K.L. (1978): Negation as failure, in: Hervé, G. & J. Miller. (eds.), *Logic and data bases, Symposium on logic and data bases, Centre d'études et de recherches de Toulouse*, Plenum Press, New York, pp. 293-322.

Clarke, S. (1998): *Metaphysics and the disunity of scientific knowledge*, Ashgate, Aldershot.

Cook, A. (1994): *The observational foundations of physics*, Cambridge University Press, Cambridge.

Creary, L.G. (1981): Causal explanation and the reality of natural component forces, *Pacific Philosophical Quarterly*, 62, pp. 148-157.

Da Costa, N.C.A. & S. French. (1990): The model-theoretic approach in the philosophy of science, *Philosophy of science*, 57, pp. 248-265.

Davidson, D. (1984): *Inquiries into truth and interpretation*, Oxford University Press, New York.

Dilworth, C. (1994): *Scientific progress. A study concerning the nature of the relation between successive theories*, Kluwer Academic Publishers, Dordrecht, 3rd ed.

Donovan, A., L. Laudan, & R. Laudan. (1988): Testing theories of scientific change, in: Donovan, A., L. Laudan, & R. Laudan. (eds.), *Scrutinising science. Empirical studies of scientific change*, Kluwer, Dordrecht, pp. 3-46.

Duhem, P. (1914): *The aim and structure of physical theory*, Princeton University Press, Princeton.

Einstein, A. (1956): *The meaning of relativity*, Princeton University Press, Princeton, 5th ed.

Feynman, R. (1965): *The character of physical law*, British Broadcasting Corporation, London.

Fine, A. (1986a): Unnatural attitudes: Realist and instrumentalist attachments to science, *Mind*, XCV(378), pp. 149-179.

Fine, A. (1986b): *The shaky game: Einstein, realism, and the quantum theory*, Chicago University Press, Chicago.

Friedman, M. (1949): The Marshallian demand curve, *Journal of Political Economy*, 57, pp. 463-495.

Friedman, M. (ed.), (1953): *Essays in positive economics*, University of Chicago Press, Chicago.

Galison, P. (1987): *How experiments end*, University of Chicago Press, Chicago.

Galison, P. (1997): *Images and logic: A material culture of microphysics*, University of Chicago Press, Chicago.

Gamov, G. (1962): *Biography of physics*, Hutchinson, London.

Gibbard, A. & H.R. Varian. (1978): Economic models, *The journal of Philosophy*, 75(11), pp. 664-677.

Giere, R.N. (1983): Testing theoretical hypotheses, in: Earman, J. (ed.), *Testing scientific theories, Minnesota studies in the philosophy of science*, X, pp. 269-298, University of Minnesota Press, Minneapolis.

Giere, R.N. (1984): Toward a unified theory of science, in: Cushing, J.T., C.F. Delaney & G. M. Gutting. (eds.), *Science and reality: Recent work in the philosophy of science*, University of Notre Dame Press, Indiana, pp. 5-31.

Giere, R.N. (1985): Constructive realism, in: Churchland, P.M. & C.A. Hooker. (eds.), *Images of science: Essays on realism and empiricism*, University of Chicago Press, Chicago, pp. 75-98.

Giere, R.N. (1991): *Understanding scientific reasoning*, Harcourt Brace Jovanovich College Publishers, New York.

Giere, R. N. (1994): The cognitive structure of scientific theories. *Philosophy of science*, 2, pp. 276-296.

Ginsberg, M.L. (ed.), (1987): *Readings in nonmonotonic reasoning*, Morgan Kaufman, California.

Grünbaum, A. (1954): Science and ideology, *The scientific monthly*, July Edition, pp. 13-19.

Haalvelmo, T. (1944): The probability approach in econometrics, *Econometrica*, 12 (supplement).

Hacking, I. (1983): *Representing and intervening. Introductory topics in the philosophy of natural science*, Cambridge University Press, Cambridge.

Hacking, I. (1997): Experimentation and scientific realism, in: Tauber, A.I. (ed.), *Science and the quest for reality*, Macmillan Press Ltd., London, pp. 162-181.

Hands, D.W. (1985): The structuralist view of economic theories: A review essay, *Economics and Philosophy*, 1, pp. 303-335.

Harrod, J. (1968): What is a model?, in: Wolfe, J. N. (ed.), *Value, capital and growth. Papers in honour of Sir John Hicks*, University Press, Edinburgh, pp. 173-191.

Hausman, D.M. (1992): *The inexact and separate science of economics*, Cambridge University Press, Cambridge.

Heidema, J. & H.J. Schutte. (1978): Truth criteria in deductive theories, *Philosophical papers*, 7, pp. 51-68.

Heidema, J. & I. Burger. (Forthcoming.): Degrees of abductive boldness.

Hempel. (1958): The theoritician's dilemma: A study in the logic of theory construction, in: Feigl, H., M. Scriven & G. Maxwell. (eds.), *Concepts, theories, and the body-mind problem. Minnesota Studies in the Philosophy of Science*, II, pp. 37-98, University of Minnesota Press, Minneapolis.

Hempel. (1973): The meaning of theoretical terms. A critique of the standard empiricist construal, in: Suppes, P., L. Henkin, J. Athanase, & Gr.C. Moisil. (eds.), *Logic, Methodology and Philosophy of Science*, IV, pp. 367-378, North-Holland, Amsterdam.

Hesse, M. (1963): *Models and analogies in science*, Oxford University Press, Oxford.

Hintikka, J. (1989): Exploring possible worlds, in: Allèn, S. (ed.), *Possible worlds in humanities, arts and sciences. Proceedings of Nobel Symposium 65*, Walter de Gruyter, Berlin.

Hintikka, J. (1997): *Lingua Universalis vs. Calculus Ratiocinator: An ultimate presupposition of the twentienth century philosophy*, Kluwer Academic Publishers, Amsterdam.

Holton, G. (1995): The role of themata in science, *Foundations in physics*, 26(4), pp. 453-465.

Hoover, K. (1994): Econometrics as observation: The Lucas Critique and the nature of econometric inference, *Journal of economic methodology*, 1, pp. 65-80.

Hutchison, T.W. (1960): *The significance and basic postulates of economic theory*, Augustus M. Kelly, New York.

Hutchison, T.W. (1977): *Knowledge and ignorance in economics*, Basil Blackwell, Oxford.

Hutchison, T.W. (1992): *Changing aims in economics*, Basil Blackwell, Oxford.

Janssen, M. (1989): Structuralist reconstructions of classical and Keynesian macroeconomics, *Erkenntnis*, 30, pp. 165-181.

Keuzenkamp, M.A. & J.R. Magnus. (1995): On tests and significance in economics, *Journal of econometrics*, 67(1), pp. 5-24.

Keynes, J.M. (1973): Defence and development, in: Moggridge, P. (ed.), *The collected writings of J.M. Keynes. The general theory and after*, XIV, MacMillan St. Martin's Press, London.

Kieseppä, I. (1996): *Truthlikeness for multidimensional, quantitative cognitive problems,* Kluwer Academic Press, Dordrecht.

Kraus, S., D. Lehmann & M. Magidor. (1990): Non-monotonic reasoning, preferential models and cumulative logics, *Artificial Inteligence*, 44, pp. 167-207.

Kripke, S.A. (1980): *Naming and necessity*, Blackwell, Oxford.

Kuhn, T.S. (1970): *The structure of scientific revolutions*, Oxford University Press, Oxford, 2nd ed.

Kuhn, T.S. (1976): Theory change as structure change: Comments on the Sneed formalism, *Erkenntnis*, 10, pp. 179-199.

Kuhn, T.S. (1977): *The essential tension. Selected studies in scientific tradition and change*, University of Chicago Press, Chicago.

Kuhn, T.S. (1997): The road since structure, in: Tauber, A.I. (ed.), *Science and the quest for reality*, Macmillan Press Ltd., London, pp. 231-248.

Kuipers, T.A.F. (1987): *What is closer-to-the-truth?* Rodopi, Amsterdam.

Kuipers, T.A.F. (1992): Naive and refined truth approximation, *Synthese*, 93, pp. 299-341.

Kuipers, T.A.F. (1994): The refined structure of theories, in: Kuokkanen, M. (ed.), *Structuralism, idealisation, and approximation, Poznán studies in the philosophy of the sciences and the humanities*, 42, pp. 3-24, Rodopi, Amsterdam.

Kuipers, T.A.F. (1997): The dual foundation of qualitative truth approximation, *Erkenntnis*, 47, pp. 145-179.

Kuipers, T.A.F. (1999): Abduction aiming at empirical progress or even truth, *Foundations of science* 4(3), pp. 307-323.

Kuipers, T.A.F. (2000): *From intrumentalism to constructive realism. On some relations between confirmation, empirical progress, and truth approximation*, Kluwer Academic Publishers, Dordrecht.

Kuipers, T.A.F. (2001): *Structures in science. Heuristic patterns based on cognitive patterns. An advanced textbook in neo-classical philosophy of science*, Synthese Library, Kluwer Academic Publishers, Dordrecht.

Lakatos, I. (1978): *The methodology of scientific research programmes*, Cambridge University Press, Cambridge.

Laudan, L. (1981a): A problem-solving approach to scientific progress, in: Hacking, I. (ed.), *Scientific revolutions*, Oxford University Press, New York, pp. 144-155.

Laudan, L. (1981b): A confutation of convergent realism, *Philosophy of science*, 48, pp. 19-49.

Laudan, L. (1984): Explaining the success of science: Beyond epistemic realism and relativism, in: Cushing, J.T., C.F. Delaney, & G.M. Gutting. (eds.), *Science and reality. Recent work in philosophy of science*, University of Notre Dame Press, Indiana, pp. 83-105.

Laudan, L. & J. Leplin. (1991): Empirical equivalence and underdetermination, *The Journal of Philosophy*, LXXXVIII(9), pp. 449-472.

Lawson, T. (1989): Abstraction, tendencies and stylised facts: A realist approach to economic analysis, *Cambridge Journal of Economics*, 13, pp. 59-78.

Lawson, T. (1994a): A realist theory for economics, in: Backhouse, R.E. (ed.), *New directions in economic methodology*, Routledge, London.

Lawson, T. (1994b): Critical realism and the analysis of choice, explanation and change, *Advances in Austrian Economics*, 1.

Lawson, T. (1997): *Economics and reality*, Routledge, London.

Laymon, R. (1989): Cartwright and the lying laws of physics, *The Journal of Philosophy*, 86, pp. 353-372.

Lenk, H. (1993): *Philosophie und Interpretation*, Suhrkamp, Frankfurt-am-Main.

Lenk, H. (1995): *Schemaspiele. Über Schemainterpretationen und Interpretationskonstrukte*, Suhrkamp, Frankfurt-am-Main.

Leplin, J. (1997): *A novel defence of scientific realism*, Oxford University Press, Oxford.

Lewis, D. (1970): How to define theoretical terms, *The Journal of Philosophy*, 67, pp. 427-446.

Lewis, D. (1984): Putnam's paradox, *The Australasian Journal of Philosophy*, 62, pp. 221-236.

Lipsey, R.G. (1983): *An introduction to positive economics*, Weidefeld & Nicolson, London, 6th ed.

Ludwig, G. (1990): *Die Grundstrukturen einer physikalischen Theorie*, Springer Verlag, Berlin, 2nd ed.

Mäki, U. (1992): Friedman and realism, *Research in the History of Economic Thought and Methodology*, 10, pp. 171-195.

Mäki, U. (1993): On the problem of realism in economics, in: Caldwell, B. (ed.), *The philosophy and methodology of economics*, Edward Elgar, Aldershot.

Mäki, U. (1994): Isolation, idealisation and truth in economics, in: *Poznan Studies in the Philosophy of the sciences and the humanities*, 38, pp. 147-168.

Mäki, U. (1996a): Two portraits of economics, *Journal of Economic Methodology*, 3, pp. 1-38.

Mäki, U. (1996b): Scientific realism and some peculiarities of economics, *Boston Studies in the Philosophy of Science*, 169, pp. 425-445.

Makwosky, J.A. (1994): The impact of model theory on computer science, in: Prawitz, D., B. Skyrms, D. Westerståhl. (eds.), *Logic, Methodology and Philosophy of Science*, IX, pp. 239-262, Elsevier, Amsterdam.

Maxwell, G. (1962): The ontological status of theoretical entities, in: Feigl, H. & G. Maxwell. (eds.), *Minnesota Studies in the Philosophy of science III*, University of Minnesota Press, Minneapolis.

Mayo, D.G. (1997): *Error and growth of experimental knowledge. (Science and its conceptual foundations)*, University of Chicago Press, Chicago.

McCarthy, J.M. (1980): Circumscription - a form of non monotonic reasoning, *Artificial Intelligence*, 13, pp. 27-39.

McCloskey, D.N. (1986): *The rhetoric of economics*, Wheatsheaf Books, Brighton.

McCloskey, D.N. (1990): *If you're so smart*, University of Chicago Press, Chicago.

McCloskey, D.N. (1994): *Rhetoric and persuasion in economics*, Cambridge University Press, Cambridge.

McDermott, D.V. (1982): A temporal logic for reasoning about processes and plans, *Cognitive Science*, 2(3), pp. 101-155.

McDermott, D.V. & J. Doyle. (1980): Non monotonic logic, *Artificial Intelligence*, 13, pp. 41-72.

McKinsey, J.C.C. & P. Suppes. (1953): Philosophy and the axiomatic foundations of physics, in: *Proceedings of the XIth International Congress of Philosophy*, 6, pp. 49-54.

McKinsey, J.C.C., A.C. Sugar, & P. Suppes. (1953): Axiomatic foundations of classic particle mechanics, *The journal of rational mechanics and analysis*, 2, pp. 253-272.

McMullin, E. (1993): Rationality and paradigm change in science, in: Horwich, P. (ed.), *World changes. T.S. Kuhn and the nature of science*, MIT Press, Cambridge, pp. 55-78.

Montague, R. M. (1962): Deterministic theories, in: Washburne, N. F. (ed.), *Decisions, values, and groups*, Pergamon Press, Oxford.

Morgan, M. (1988): Finding a satisfactory empirical model, in: De Marchi, N. (ed.), *The Popperian legacy in economics*, Cambridge University Press, Cambridge, pp. 199-211.

Morgan, M. (1990): *The history of econometric ideas*, Cambridge University Press, Cambridge.

Morgan, M. & M. Morrison. (eds.), (2000): *Models as mediators: Perspectives on natural and social science. (Ideas in context)*, Cambridge University Press, Cambridge.

Moulines, C.U. (1991): Pragmatics in the structuralist view of science, in: Schurz, G. & G. J. W. Dorn. (eds.), *Advances in scientific philosophy. Essays in honour of Paul Weingartner*, Rodopi, Amsterdam, pp. 313-326.

Musgrave, A. (1981): On interpreting Friedman, *KYKLOS*, (34), pp. 377-387.

Nagel, E. (1961): *The structure of science*, Routledge & Kegan Paul, London.

Nagel. E. (1963): Wholes, sums and organic unities, in: Lerner, D. (ed.), *Parts and wholes*, The Free Press, New York.

Nersessian, N.J. (1984): *Faraday to Einstein: Constructing meaning in scientific theories*, Martinus Nijhoff Publishers, Dordrecht.

Newton, I. (1952): *Opticks, or, A treatise of the reflections, refractions, inflections, and colours of light*, Dover Publications, London, based on the 4th ed., London, 1730.

Niiniluoto, I. (1984): *Is science progressive?*, D. Reidel, Dordrecht.

Niiniluoto, I. (1987): *Truthlikeness*, D. Reidel, Dordrecht.

Niiniluoto, I. (1998): Verisimilitude: The third period. (Survey Article), *The British journal for the philosophy of science*, 49, pp. 1-29.

Niiniluoto, I. (1999): *Critical scientific realism*, Oxford University Press, Oxford.

Norris, C. (1997): *Against relativism. Philosophy of science, deconstruction and critical theory*, Blackwell, Oxford.

Oddie, G. (1986): *Likeness to the truth*, D. Reidel, Dordrecht.

Papineau, D. (1995): Philosophy of science, in: Honderich, T. (ed.), *The Oxford companion to philosophy*, Oxford University Press, New York, pp. 809-812.

Patinkin, D. (1965): *Money, interest and prices. An integration of monetary and value theory*, Harper and Row, London.

Paul, G. (1993): Approaches to abductive reasoning: An overview, *Artificial Intelligence Review* 7, pp.109-152.

Pearce, D. & V. Rantala. (1983): New foundations for metascience, *Synthese*, 56, pp. 1-26.

Penrose, R. (1997): The mysteries of quantum physics, in: Penrose, R. (ed.), *The large, the small and the human mind*, Cambridge University Press, Cambridge, pp. 50-92.

Popper, K.R. (1989): *Conjectures and refutations. The growth of scientific knowledge*, Routledge, London, 5th ed. revised.

Popper, K.R. (1990): *A world of propensities*, Thoemmes, Bristol.

Przelecki, M. (1969): *The logic of empirical theories*, Routledge & Kegan Paul, London.

Przelecki, M. (1974): On a model-theoretic approach to empirical interpretation of scientific theories, *Synthese*, 26, pp. 401- 406.

Przelecki, M. (1991): Is the notion of truth applicable to scientific theories?, in: Schurz, G. & G.J.W. Dorn. (eds.), *Advances in scientific philosophy. Essays in honour of Paul Weingartner*, Rodopi, Amsterdam, pp. 283-294.

Przelecki, M. & R. Wójcicki. (1969): The problem of analiticity, *Synthese*, 18, pp. 374-399.

Psillos, S. (1999a): "How not to structure realism", read at the LMPS conference in Cracow in 1999.

Psillos, S. (1999b): *Scientific realism: How science tracks truth (Philosophical issues in science)*, Routledge, Boston.

Putnam, H. (1962): What theories are not, in: Nagel, E., P. Suppes & A. Tarski. (eds.), *Logic, Methodology, and Philosophy of Science: Proceedings of the 1960 international congress*, Stanford University Press, Stanford.

Putnam, H. (1975): *Mathematics, matter and methodology*, Cambridge University Press, Cambridge.

Putnam, H. (1978): Realism and reason, in: Putnam, H. *Meaning and the moral sciences*, Routledge & Kegan Paul, London, pp. 123-140.

Putnam, H.. (1983): Models and reality, in: Putnam, H. *Realism and reason*, Cambridge University Press, New York, pp. 1-25.

Rantala, V. (1978): The old and the new logic of metascience, *Synthese*, 39, pp. 233-247.

Rantala, V. (1980): On the logical basis of the structuralist philosophy of science, *Erkenntnis*, 15, pp. 269-286.

Rappaport, S. (1988): Arguments, truth and economic methodology: A rejoinder to McCloskey, *Economics and Philosophy*, 4, pp. 170-172.

Rappaport, S. (1993): Must a metaphysical relativist be a truth relativist? *Philosophia*, 22, pp. 75-85.

Rappaport, S. (1996): Inference to the best explanation; is it really different from Mill's methods?, *Philosophy of Science*, 63, pp. 65-80.

Rappaport, S. (1998): *Models and reality in economics*, Edward Elgar, Cheltenham.

Redhead, M. L. G. (1980): Models in physics, *The British journal for the philosophy of science*, 31, pp. 145-163.

Reiter, R. (1980): A logic for default reasoning, *Artificial Intelligence*, 13, pp. 81-132.

Robinson, A. (1986): *Introduction to model-theory and the metamathematics of algebra*, North Holland Publishers, New York, 2nd ed.

Rosenberg, A. (1985): Methodology, theory and the philosophy of science, *Pacific Philosophical Quarterly*, 66, pp. 377-393.

Rosenberg, A. (1986): Lakatosian consolations for economists, *Economics and Philosophy*, 2, pp. 127-139.

Ross, D. (1999): Folk theories, models and economic reality. A reply to Williams, *The South African journal of philosophy*, 18(2), pp. 247-257.

Rubin, H. & P. Suppes. (1954): Transformations of systems of relativistic particle mechanics, *Pacific journal of mathematics*, 4, pp. 563-601.

Rueger, A. & W.D. Sharp. (1996): Simple theories of a messy world: Truth and explanatory power in nonlinear dynamics, *The British journal for the philosophy of science*, 47, pp. 93-112.

Russell, B. (1903): *The principles of mathematics*, Cambridge University Press, Cambridge.

Ruttkamp, E.B. (1997a): The role of models in philosophy of science: Mediating between the "general" and the "particular", in: Forrai, G. (ed.), *Images and reality. Proceedings of the 1996 Miskolc conference*, Miskolci Egyetem Tár, Miskolc, pp. 127-138.

Ruttkamp, E.B. (1997b): A model-theoretic interpretation of science, *The South African journal of philosophy*, 16(1), pp. 31-36.

Ruttkamp, E.B. (1999a): Semantic approaches in the philosophy of science, *The South African journal of philosophy, Special Issue on Philosophy of Science*, 18(2), pp.100-148.

Ruttkamp, E.B. 1999b. Reality in science, *The South African journal of philosophy, Special Issue on Philosophy of Science*, 18(2), pp. 149-191.

Ryan, M. & P.-Y. Schobbens. (1995): Belief revision and verisimilitude, *The Notre Dame journal of formal logic*, 36, pp. 15-29.

Sachs, M. (1988): *Einstein versus Bohr*, Open Court Publishing Company, Illinois.

Sarlemijn, A. & M. Sparnaay. (eds.), (1989): *Physics in the making. Essays on developments in 20th century physics. In honour of H.B.G. Casimir*, North Holland Publishers, New York.

Sartori, L. (1996): *Understanding relativity. A simplified approach to Einstein's theories*, California University Press, Berkeley.

Schurz, G. (1995): Theories and their applications - a case of nonmonotonic reasoning, in: Herfel, W.E., W. Krajewski, I. Niiniluoto & R. Wójcicki. (eds.), *Theories and models in scientific processes. Poznán studies in the philosophy of sciences and the humanities*, 44, pp. 269-294, Rodopi, Amsterdam.

Schwinger, J. (1986): *Einstein's legacy*, Scientific American Books, New York.

Scott, D. & P. Suppes. (1958): Foundational aspects of theories of measurement, *The journal of symbolic logic*, 23, pp. 113-128.

Scriven, M. (1959): Truisms as grounds for historical explanations, in: Gardiner, P. (ed.), Theories of history, Free Press, New York, pp. 443-468.

Sellars, W. (1957): Counterfactuals, dispositions, and the causal modalities, in: *Minnesota studies in the philosophy of science*, II, pp. 225-308, University of Minnesota Press, Minneapolis.

Sellars, W. (1963): *Science, perception, and reality*, Humanities Press, New York.

Shoham, Y. (1987): A semantical approach to nonmonotonic logics, in: *Proceedings: Logics in Computer Science*, pp. 275-279.

Shoham, Y. (1988): *Reasoning about change: time and causation from the standpoint of artificial intelligence*, MIT Press, Massachusetts.

Sneed, J.D. (1976): Philosophical problems in the empirical science of science: a formal approach, *Erkenntnis*, 10, pp. 115-146.

Sneed, J.D. (1979): *The logical structure of mathematical physics*, D. Reidel, Dordrecht, 2nd ed. revised.

Sneed, J.D. (1983): Structuralism and scientific realism, *Erkenntnis*, 19, pp. 245-370.

Sneed, J.D. (1994): Structural explanation, in: Humphreys, P. (ed.), *Patrick Suppes: Scientific philosopher. Volume 2, Philosophy of physics, theory structure, and measurement theory*, Kluwer Academic Publishers, Dordrecht, pp. 195-216.

Spurrett, D. (1998): Transcendental realism defended: a response to Allan, *The South African journal of Philosophy*, 17(3), pp. 198-210.

Spurret, D. (1999): Cartwright on laws and composition, *International Studies in the Philosophy of Science*.

Stegmüller, W. (1976): *The structure and dynamics of theories*, Springer-Verlag, Berlin.

Stegmüller, W. (1979): *The structuralist view of theories: A possible analogue of the Bourbaki programme in physical science*, Springer-Verlag, Berlin.

Stegmüller, W. (1979): The structuralist view: Survey, recent developments and answers to some criticisms, in: Hintikka, J. (ed.), *The logic and epistemology of scientific change. Acta Philosophica Fennica*, **XXX**, pp. 113-129, North-Holland Publishing Company, Amsterdam.

Stegmüller, W. (1986): *Probleme und Resultate der Wissenschaftstheorie und analytischen Philosophie. Band II: Dritter Teilband: Die Entwicklung des neuen Strukturalismus seit 1973*, Springer verlag, Berlin.

Suchting, W. (1992): Reflections upon Roy Bhaskar's "critical realism", *Radical Philosophy*, 61, pp. 23-31.

Suppe, F. (1967): *The meaning and uses of models in mathematics and the exact sciences*, Ph.D. Thesis, University of Michigan.

Suppe, F. (1973): Theories, their formulations and the operational imperative, *Synthese*, 25, pp. 129-164.

Suppe, F. (1989): *The semantic conception of theories and scientific realism*, University of Illinois Press, Illinois.

Suppes, P. (1954): Some remarks on problems and methods in the philosophy of science, *Philosophy of science*, 21, pp. 242-248.

Suppes, P. (1959): Axioms for relativistic kinematics with or without parity, in: Henkin, L., P. Suppes, & A. Tarski. (eds.), *The axiomatic method*, North-Holland Publishers, Amsterdam.

Suppes. P. (1960): A comparison of the meaning and uses of models in mathematics and the empirical sciences, *Synthese*, 12, pp. 287-300.

Suppes, P. (1967): What is a scientific theory?, in: Morgenbesser, S. (ed.), *Philosophy of science today*, Basic Books, New York, pp. 55-67.

Suppes, P. (1969): *Studies in the methodology and the foundations of science*, D. Reidel & Co, Dordrecht.

Suppes, P. (1988a): Philosophical implications of Tarski's work, *The journal of symbolic logic*, 53, pp. 80-91.

Suppes, P. (1988b): Representation theory and the analysis of structure, *Philosophia Naturalis*, 25, pp. 254-268.

Suppes, P. (1989): *Studies in the methodology and foundations of science. Selected papers from 1951 to 1969*, D. Reidel, Dordrecht, Part I, pp. 1-80.

Suppes, P. (1993): *Models and methods in the philosophy of science: Selected essays*, Kluwer Academic Publishers, Dordrecht.

Tarski, A. (1935): Der Wahrheitsbegriff in den formalisierten Sprachen, *Studia Philosophica*, 1, pp. 261-405. (English translation by J.H. Woodger. (1956): The concept of truth in formalised languages, in: *Logic, semantics, metamathematics*, Oxford University Press, Oxford, pp. 152-278.)

Torr, C. (1999): Equilibrium and equilibrium, in: Earl, P.E. et al. (eds.), *Contingency, complexity and the theory of the firm*, II. Edward Elgar, Cheltenham.

Torretti, R. (1990): *Creative understanding: Philosophical reflections on physics*, University of Chicago Press, Chicago.

Tuomela, R. (1972a): Deductive explanation of scientific laws, *The journal of philosophical logic*, 1, pp. 369-392.

Tuomela, R. (1972b): Model theory and empirical interpretation of scientific theories, *Synthese*, 25, pp. 165-175.

Tuomela, R. (1974): Empiricist vs realist semantics and model theory, *Synthese*, 26, pp. 407-408.

Van Fraassen, B.C. (1969): Meaning relations among predicates, *NOÛS*, 3, pp. 155-167.

Van Fraassen, B.C. (1970): On the extension of Beth's semantics for physical theories, *Philosophy of science*, 37, pp. 325-339.

Van Fraassen, B.C. (1976): To save the phenomena, *The Journal of Philosophy*, LXXIII(18), pp. 623-632.

Van Fraassen, B. C. (1980): *The scientific image*, Oxford University Press, Oxford.

Van Fraassen, B.C. (1994): The world of empiricism, in: Hilgevoord, J. (ed.), *Physics and our view of the world*, Cambridge Universitty Press, Cambridge, pp. 114-134.

Van Fraassen, B.C. (1997): Putnam's paradox: metaphysical realism revamped and evaded, in: Tomberlin, J.E. (ed.), *Philosophical perspectives, 11, Mind, Causation, and World, A supplement to NOÛS*, Blackwell Publishers, Boston, pp. 17-42.

Von Neumann, J. (1955): *Mathematical foundations of quantum mechanics*, Princeton University Press, Princeton.

Wartofsky, M. W. (1979): *Models: Representation and understanding of science*, D. Reidel, Dordrecht.

Weintraub. E.R. (1974): *General equilibrium theory*, MacMillan, London.

Weintraub, E.R. (1992): Roger Backhouse's sraw herring, *Methodus*, 4(2), pp. 53-57.

Wójcicki, R. (1979): *Topics in the formal methodology of empirical sciences*, D. Reidel, Dordrecht.

Wójcicki, R. (1994): Theories and theoretical models, in: Humphreys, P. (ed.), *Patrick Suppes: Scientific philosopher. Volume 2, Philosophy of physics, theory structure, and measurement theory*, Kluwer Academic Publishers, Dordrecht, pp. 125-149.

Wójcicki, R. (1996): Theories in science, in: Bystrov, P.I. & V.N. Sadovsky. (eds.), *Philosophical logic and logical philosophy. Essays in honour of Vladimir A. Smirnov*, Kluwer Academic Press, Dordrecht.

Worrall, J. (1994): Introduction, in: Worrall, J. (ed.), *The ontology of science*, Dartmouth, Aldershot, pp. xi-xxx.

Zamora Bonilla, J.P. (1996): Verisimilitude, structuralism, and scientific progress, *Erkenntnis*, 44, pp. 25-48.

Zwart, S.D. (1998): *Approach to the truth. Verisimilitude and truthlikeness*, Institute for Logic, Language and Computation, Amsterdam.

SYNTHESE LIBRARY

1. J. M. Bochénski, *A Precis of Mathematical Logic*. Translated from French and German by O. Bird. 1959 ISBN 90-277-0073-7
2. P. Guiraud, *Problèmes et méthodes de la statistique linguistique*. 1959 ISBN 90-277-0025-7
3. H. Freudenthal (ed.), *The Concept and the Role of the Model in Mathematics and Natural and Social Sciences*. 1961 ISBN 90-277-0017-6
4. E. W. Beth, *Formal Methods*. An Introduction to Symbolic Logic and to the Study of Effective Operations in Arithmetic and Logic. 1962 ISBN 90-277-0069-9
5. B. H. Kazemier and D. Vuysje (eds.), *Logic and Language*. Studies dedicated to Professor Rudolf Carnap on the Occasion of His 70th Birthday. 1962 ISBN 90-277-0019-2
6. M. W. Wartofsky (ed.), *Proceedings of the Boston Colloquium for the Philosophy of Science, 1961–1962*. [Boston Studies in the Philosophy of Science, Vol. I] 1963 ISBN 90-277-0021-4
7. A. A. Zinov'ev, *Philosophical Problems of Many-valued Logic*. A revised edition, edited and translated (from Russian) by G. Küng and D.D. Comey. 1963 ISBN 90-277-0091-5
8. G. Gurvitch, *The Spectrum of Social Time*. Translated from French and edited by M. Korenbaum and P. Bosserman. 1964 ISBN 90-277-0006-0
9. P. Lorenzen, *Formal Logic*. Translated from German by F.J. Crosson. 1965 ISBN 90-277-0080-X
10. R. S. Cohen and M. W. Wartofsky (eds.), *Proceedings of the Boston Colloquium for the Philosophy of Science, 1962–1964*. In Honor of Philipp Frank. [Boston Studies in the Philosophy of Science, Vol. II] 1965 ISBN 90-277-9004-0
11. E. W. Beth, *Mathematical Thought*. An Introduction to the Philosophy of Mathematics. 1965 ISBN 90-277-0070-2
12. E. W. Beth and J. Piaget, *Mathematical Epistemology and Psychology*. Translated from French by W. Mays. 1966 ISBN 90-277-0071-0
13. G. Küng, *Ontology and the Logistic Analysis of Language*. An Enquiry into the Contemporary Views on Universals. Revised ed., translated from German. 1967 ISBN 90-277-0028-1
14. R. S. Cohen and M. W. Wartofsky (eds.), *Proceedings of the Boston Colloquium for the Philosophy of Sciences, 1964–1966*. In Memory of Norwood Russell Hanson. [Boston Studies in the Philosophy of Science, Vol. III] 1967 ISBN 90-277-0013-3
15. C. D. Broad, *Induction, Probability, and Causation*. Selected Papers. 1968 ISBN 90-277-0012-5
16. G. Patzig, *Aristotle's Theory of the Syllogism*. A Logical-philosophical Study of *Book A* of the *Prior Analytics*. Translated from German by J. Barnes. 1968 ISBN 90-277-0030-3
17. N. Rescher, *Topics in Philosophical Logic*. 1968 ISBN 90-277-0084-2
18. R. S. Cohen and M. W. Wartofsky (eds.), *Proceedings of the Boston Colloquium for the Philosophy of Science, 1966–1968, Part I*. [Boston Studies in the Philosophy of Science, Vol. IV] 1969 ISBN 90-277-0014-1
19. R. S. Cohen and M. W. Wartofsky (eds.), *Proceedings of the Boston Colloquium for the Philosophy of Science, 1966–1968, Part II*. [Boston Studies in the Philosophy of Science, Vol. V] 1969 ISBN 90-277-0015-X
20. J. W. Davis, D. J. Hockney and W. K. Wilson (eds.), *Philosophical Logic*. 1969 ISBN 90-277-0075-3
21. D. Davidson and J. Hintikka (eds.), *Words and Objections*. Essays on the Work of W. V. Quine. 1969, rev. ed. 1975 ISBN 90-277-0074-5; Pb 90-277-0602-6
22. P. Suppes, *Studies in the Methodology and Foundations of Science. Selected Papers from 1951 to 1969*. 1969 ISBN 90-277-0020-6
23. J. Hintikka, *Models for Modalities*. Selected Essays. 1969 ISBN 90-277-0078-8; Pb 90-277-0598-4

SYNTHESE LIBRARY

24. N. Rescher *et al.* (eds.), *Essays in Honor of Carl G. Hempel*. A Tribute on the Occasion of His 65th Birthday. 1969　ISBN 90-277-0085-0
25. P. V. Tavanec (ed.), *Problems of the Logic of Scientific Knowledge*. Translated from Russian. 1970　ISBN 90-277-0087-7
26. M. Swain (ed.), *Induction, Acceptance, and Rational Belief*. 1970　ISBN 90-277-0086-9
27. R. S. Cohen and R. J. Seeger (eds.), *Ernst Mach: Physicist and Philosopher*. [Boston Studies in the Philosophy of Science, Vol. VI]. 1970　ISBN 90-277-0016-8
28. J. Hintikka and P. Suppes, *Information and Inference*. 1970　ISBN 90-277-0155-5
29. K. Lambert, *Philosophical Problems in Logic*. Some Recent Developments. 1970　ISBN 90-277-0079-6
30. R. A. Eberle, *Nominalistic Systems*. 1970　ISBN 90-277-0161-X
31. P. Weingartner and G. Zecha (eds.), *Induction, Physics, and Ethics*. 1970 ISBN 90-277-0158-X
32. E. W. Beth, *Aspects of Modern Logic*. Translated from Dutch. 1970　ISBN 90-277-0173-3
33. R. Hilpinen (ed.), *Deontic Logic*. Introductory and Systematic Readings. 1971 *See also* No. 152.　ISBN Pb (1981 rev.) 90-277-1302-2
34. J.-L. Krivine, *Introduction to Axiomatic Set Theory*. Translated from French. 1971　ISBN 90-277-0169-5; Pb 90-277-0411-2
35. J. D. Sneed, *The Logical Structure of Mathematical Physics*. 2nd rev. ed., 1979　ISBN 90-277-1056-2; Pb 90-277-1059-7
36. C. R. Kordig, *The Justification of Scientific Change*. 1971　ISBN 90-277-0181-4; Pb 90-277-0475-9
37. M. Čapek, *Bergson and Modern Physics*. A Reinterpretation and Re-evaluation. [Boston Studies in the Philosophy of Science, Vol. VII] 1971　ISBN 90-277-0186-5
38. N. R. Hanson, *What I Do Not Believe, and Other Essays*. Ed. by S. Toulmin and H. Woolf. 1971　ISBN 90-277-0191-1
39. R. C. Buck and R. S. Cohen (eds.), *PSA 1970*. Proceedings of the Second Biennial Meeting of the Philosophy of Science Association, Boston, Fall 1970. In Memory of Rudolf Carnap. [Boston Studies in the Philosophy of Science, Vol. VIII] 1971　ISBN 90-277-0187-3; Pb 90-277-0309-4
40. D. Davidson and G. Harman (eds.), *Semantics of Natural Language*. 1972　ISBN 90-277-0304-3; Pb 90-277-0310-8
41. Y. Bar-Hillel (ed.), *Pragmatics of Natural Languages*. 1971　ISBN 90-277-0194-6; Pb 90-277-0599-2
42. S. Stenlund, *Combinators, γ Terms and Proof Theory*. 1972　ISBN 90-277-0305-1
43. M. Strauss, *Modern Physics and Its Philosophy*. Selected Paper in the Logic, History, and Philosophy of Science. 1972　ISBN 90-277-0230-6
44. M. Bunge, *Method, Model and Matter*. 1973　ISBN 90-277-0252-7
45. M. Bunge, *Philosophy of Physics*. 1973　ISBN 90-277-0253-5
46. A. A. Zinov'ev, *Foundations of the Logical Theory of Scientific Knowledge (Complex Logic)*. Revised and enlarged English edition with an appendix by G. A. Smirnov, E. A. Sidorenka, A. M. Fedina and L. A. Bobrova. [Boston Studies in the Philosophy of Science, Vol. IX] 1973　ISBN 90-277-0193-8; Pb 90-277-0324-8
47. L. Tondl, *Scientific Procedures*. A Contribution concerning the Methodological Problems of Scientific Concepts and Scientific Explanation. Translated from Czech by D. Short. Edited by R.S. Cohen and M.W. Wartofsky. [Boston Studies in the Philosophy of Science, Vol. X] 1973　ISBN 90-277-0147-4; Pb 90-277-0323-X
48. N. R. Hanson, *Constellations and Conjectures*. 1973　ISBN 90-277-0192-X

SYNTHESE LIBRARY

49. K. J. J. Hintikka, J. M. E. Moravcsik and P. Suppes (eds.), *Approaches to Natural Language*. 1973   ISBN 90-277-0220-9; Pb 90-277-0233-0
50. M. Bunge (ed.), *Exact Philosophy*. Problems, Tools and Goals. 1973   ISBN 90-277-0251-9
51. R. J. Bogdan and I. Niiniluoto (eds.), *Logic, Language and Probability*. 1973   ISBN 90-277-0312-4
52. G. Pearce and P. Maynard (eds.), *Conceptual Change*. 1973   ISBN 90-277-0287-X; Pb 90-277-0339-6
53. I. Niiniluoto and R. Tuomela, *Theoretical Concepts and Hypothetico-inductive Inference*. 1973   ISBN 90-277-0343-4
54. R. Fraissé, *Course of Mathematical Logic* – Volume 1: *Relation and Logical Formula*. Translated from French. 1973   ISBN 90-277-0268-3; Pb 90-277-0403-1
(For *Volume 2* see under No. 69).
55. A. Grünbaum, *Philosophical Problems of Space and Time*. Edited by R.S. Cohen and M.W. Wartofsky. 2nd enlarged ed. [Boston Studies in the Philosophy of Science, Vol. XII] 1973   ISBN 90-277-0357-4; Pb 90-277-0358-2
56. P. Suppes (ed.), *Space, Time and Geometry*. 1973   ISBN 90-277-0386-8; Pb 90-277-0442-2
57. H. Kelsen, *Essays in Legal and Moral Philosophy*. Selected and introduced by O. Weinberger. Translated from German by P. Heath. 1973   ISBN 90-277-0388-4
58. R. J. Seeger and R. S. Cohen (eds.), *Philosophical Foundations of Science*. [Boston Studies in the Philosophy of Science, Vol. XI] 1974   ISBN 90-277-0390-6; Pb 90-277-0376-0
59. R. S. Cohen and M. W. Wartofsky (eds.), *Logical and Epistemological Studies in Contemporary Physics*. [Boston Studies in the Philosophy of Science, Vol. XIII] 1973   ISBN 90-277-0391-4; Pb 90-277-0377-9
60. R. S. Cohen and M. W. Wartofsky (eds.), *Methodological and Historical Essays in the Natural and Social Sciences*. Proceedings of the Boston Colloquium for the Philosophy of Science, 1969–1972. [Boston Studies in the Philosophy of Science, Vol. XIV] 1974   ISBN 90-277-0392-2; Pb 90-277-0378-7
61. R. S. Cohen, J. J. Stachel and M. W. Wartofsky (eds.), *For Dirk Struik. Scientific, Historical and Political Essays*. [Boston Studies in the Philosophy of Science, Vol. XV] 1974   ISBN 90-277-0393-0; Pb 90-277-0379-5
62. K. Ajdukiewicz, *Pragmatic Logic*. Translated from Polish by O. Wojtasiewicz. 1974   ISBN 90-277-0326-4
63. S. Stenlund (ed.), *Logical Theory and Semantic Analysis*. Essays dedicated to Stig Kanger on His 50th Birthday. 1974   ISBN 90-277-0438-4
64. K. F. Schaffner and R. S. Cohen (eds.), *PSA 1972. Proceedings of the Third Biennial Meeting of the Philosophy of Science Association*. [Boston Studies in the Philosophy of Science, Vol. XX] 1974   ISBN 90-277-0408-2; Pb 90-277-0409-0
65. H. E. Kyburg, Jr., *The Logical Foundations of Statistical Inference*. 1974   ISBN 90-277-0330-2; Pb 90-277-0430-9
66. M. Grene, *The Understanding of Nature*. Essays in the Philosophy of Biology. [Boston Studies in the Philosophy of Science, Vol. XXIII] 1974   ISBN 90-277-0462-7; Pb 90-277-0463-5
67. J. M. Broekman, *Structuralism: Moscow, Prague, Paris*. Translated from German. 1974   ISBN 90-277-0478-3
68. N. Geschwind, *Selected Papers on Language and the Brain*. [Boston Studies in the Philosophy of Science, Vol. XVI] 1974   ISBN 90-277-0262-4; Pb 90-277-0263-2
69. R. Fraissé, *Course of Mathematical Logic* – Volume 2: *Model Theory*. Translated from French. 1974   ISBN 90-277-0269-1; Pb 90-277-0510-0
(For *Volume 1* see under No. 54)

SYNTHESE LIBRARY

70. A. Grzegorczyk, *An Outline of Mathematical Logic*. Fundamental Results and Notions explained with all Details. Translated from Polish. 1974   ISBN 90-277-0359-0; Pb 90-277-0447-3
71. F. von Kutschera, *Philosophy of Language*. 1975   ISBN 90-277-0591-7
72. J. Manninen and R. Tuomela (eds.), *Essays on Explanation and Understanding*. Studies in the Foundations of Humanities and Social Sciences. 1976   ISBN 90-277-0592-5
73. J. Hintikka (ed.), *Rudolf Carnap, Logical Empiricist*. Materials and Perspectives. 1975   ISBN 90-277-0583-6
74. M. Čapek (ed.), *The Concepts of Space and Time*. Their Structure and Their Development. [Boston Studies in the Philosophy of Science, Vol. XXII] 1976   ISBN 90-277-0355-8; Pb 90-277-0375-2
75. J. Hintikka and U. Remes, *The Method of Analysis*. Its Geometrical Origin and Its General Significance. [Boston Studies in the Philosophy of Science, Vol. XXV] 1974   ISBN 90-277-0532-1; Pb 90-277-0543-7
76. J. E. Murdoch and E. D. Sylla (eds.), *The Cultural Context of Medieval Learning*. [Boston Studies in the Philosophy of Science, Vol. XXVI] 1975   ISBN 90-277-0560-7; Pb 90-277-0587-9
77. S. Amsterdamski, *Between Experience and Metaphysics*. Philosophical Problems of the Evolution of Science. [Boston Studies in the Philosophy of Science, Vol. XXXV] 1975   ISBN 90-277-0568-2; Pb 90-277-0580-1
78. P. Suppes (ed.), *Logic and Probability in Quantum Mechanics*. 1976   ISBN 90-277-0570-4; Pb 90-277-1200-X
79. H. von Helmholtz: *Epistemological Writings. The Paul Hertz / Moritz Schlick Centenary Edition of 1921 with Notes and Commentary by the Editors*. Newly translated from German by M. F. Lowe. Edited, with an Introduction and Bibliography, by R. S. Cohen and Y. Elkana. [Boston Studies in the Philosophy of Science, Vol. XXXVII] 1975   ISBN 90-277-0290-X; Pb 90-277-0582-8
80. J. Agassi, *Science in Flux*. [Boston Studies in the Philosophy of Science, Vol. XXVIII] 1975   ISBN 90-277-0584-4; Pb 90-277-0612-2
81. S. G. Harding (ed.), *Can Theories Be Refuted?* Essays on the Duhem-Quine Thesis. 1976   ISBN 90-277-0629-8; Pb 90-277-0630-1
82. S. Nowak, *Methodology of Sociological Research*. General Problems. 1977   ISBN 90-277-0486-4
83. J. Piaget, J.-B. Grize, A. Szemińsska and V. Bang, *Epistemology and Psychology of Functions*. Translated from French. 1977   ISBN 90-277-0804-5
84. M. Grene and E. Mendelsohn (eds.), *Topics in the Philosophy of Biology*. [Boston Studies in the Philosophy of Science, Vol. XXVII] 1976   ISBN 90-277-0595-X; Pb 90-277-0596-8
85. E. Fischbein, *The Intuitive Sources of Probabilistic Thinking in Children*. 1975   ISBN 90-277-0626-3; Pb 90-277-1190-9
86. E. W. Adams, *The Logic of Conditionals*. An Application of Probability to Deductive Logic. 1975   ISBN 90-277-0631-X
87. M. Przełęcki and R. Wójcicki (eds.), *Twenty-Five Years of Logical Methodology in Poland*. Translated from Polish. 1976   ISBN 90-277-0601-8
88. J. Topolski, *The Methodology of History*. Translated from Polish by O. Wojtasiewicz. 1976   ISBN 90-277-0550-X
89. A. Kasher (ed.), *Language in Focus: Foundations, Methods and Systems*. Essays dedicated to Yehoshua Bar-Hillel. [Boston Studies in the Philosophy of Science, Vol. XLIII] 1976   ISBN 90-277-0644-1; Pb 90-277-0645-X

SYNTHESE LIBRARY

90. J. Hintikka, *The Intentions of Intentionality and Other New Models for Modalities.* 1975
    ISBN 90-277-0633-6; Pb 90-277-0634-4
91. W. Stegmüller, *Collected Papers on Epistemology, Philosophy of Science and History of Philosophy.* 2 Volumes. 1977   Set ISBN 90-277-0767-7
92. D. M. Gabbay, *Investigations in Modal and Tense Logics with Applications to Problems in Philosophy and Linguistics.* 1976   ISBN 90-277-0656-5
93. R. J. Bogdan, *Local Induction.* 1976   ISBN 90-277-0649-2
94. S. Nowak, *Understanding and Prediction.* Essays in the Methodology of Social and Behavioral Theories. 1976   ISBN 90-277-0558-5; Pb 90-277-1199-2
95. P. Mittelstaedt, *Philosophical Problems of Modern Physics.* [Boston Studies in the Philosophy of Science, Vol. XVIII] 1976   ISBN 90-277-0285-3; Pb 90-277-0506-2
96. G. Holton and W. A. Blanpied (eds.), *Science and Its Public: The Changing Relationship.* [Boston Studies in the Philosophy of Science, Vol. XXXIII] 1976
    ISBN 90-277-0657-3; Pb 90-277-0658-1
97. M. Brand and D. Walton (eds.), *Action Theory.* 1976   ISBN 90-277-0671-9
98. P. Gochet, *Outline of a Nominalist Theory of Propositions.* An Essay in the Theory of Meaning and in the Philosophy of Logic. 1980   ISBN 90-277-1031-7
99. R. S. Cohen, P. K. Feyerabend, and M. W. Wartofsky (eds.), *Essays in Memory of Imre Lakatos.* [Boston Studies in the Philosophy of Science, Vol. XXXIX] 1976
    ISBN 90-277-0654-9; Pb 90-277-0655-7
100. R. S. Cohen and J. J. Stachel (eds.), *Selected Papers of Léon Rosenfield.* [Boston Studies in the Philosophy of Science, Vol. XXI] 1979   ISBN 90-277-0651-4; Pb 90-277-0652-2
101. R. S. Cohen, C. A. Hooker, A. C. Michalos and J. W. van Evra (eds.), *PSA 1974. Proceedings of the 1974 Biennial Meeting of the Philosophy of Science Association.* [Boston Studies in the Philosophy of Science, Vol. XXXII] 1976   ISBN 90-277-0647-6; Pb 90-277-0648-4
102. Y. Fried and J. Agassi, *Paranoia.* A Study in Diagnosis. [Boston Studies in the Philosophy of Science, Vol. L] 1976   ISBN 90-277-0704-9; Pb 90-277-0705-7
103. M. Przełęcki, K. Szaniawski and R. Wójcicki (eds.), *Formal Methods in the Methodology of Empirical Sciences.* 1976   ISBN 90-277-0698-0
104. J. M. Vickers, *Belief and Probability.* 1976   ISBN 90-277-0744-8
105. K. H. Wolff, *Surrender and Catch.* Experience and Inquiry Today. [Boston Studies in the Philosophy of Science, Vol. LI] 1976   ISBN 90-277-0758-8; Pb 90-277-0765-0
106. K. Kosík, *Dialectics of the Concrete.* A Study on Problems of Man and World. [Boston Studies in the Philosophy of Science, Vol. LII] 1976   ISBN 90-277-0761-8; Pb 90-277-0764-2
107. N. Goodman, *The Structure of Appearance.* 3rd ed. with an Introduction by G. Hellman. [Boston Studies in the Philosophy of Science, Vol. LIII] 1977
    ISBN 90-277-0773-1; Pb 90-277-0774-X
108. K. Ajdukiewicz, *The Scientific World-Perspective and Other Essays, 1931-1963.* Translated from Polish. Edited and with an Introduction by J. Giedymin. 1978   ISBN 90-277-0527-5
109. R. L. Causey, *Unity of Science.* 1977   ISBN 90-277-0779-0
110. R. E. Grandy, *Advanced Logic for Applications.* 1977   ISBN 90-277-0781-2
111. R. P. McArthur, *Tense Logic.* 1976   ISBN 90-277-0697-2
112. L. Lindahl, *Position and Change.* A Study in Law and Logic. Translated from Swedish by P. Needham. 1977   ISBN 90-277-0787-1
113. R. Tuomela, *Dispositions.* 1978   ISBN 90-277-0810-X
114. H. A. Simon, *Models of Discovery and Other Topics in the Methods of Science.* [Boston Studies in the Philosophy of Science, Vol. LIV] 1977   ISBN 90-277-0812-6; Pb 90-277-0858-4

SYNTHESE LIBRARY

115. R. D. Rosenkrantz, *Inference, Method and Decision*. Towards a Bayesian Philosophy of Science. 1977 ISBN 90-277-0817-7; Pb 90-277-0818-5
116. R. Tuomela, *Human Action and Its Explanation*. A Study on the Philosophical Foundations of Psychology. 1977 ISBN 90-277-0824-X
117. M. Lazerowitz, *The Language of Philosophy*. Freud and Wittgenstein. [Boston Studies in the Philosophy of Science, Vol. LV] 1977 ISBN 90-277-0826-6; Pb 90-277-0862-2
118. Not published 119. J. Pelc (ed.), *Semiotics in Poland, 1894–1969*. Translated from Polish. 1979 ISBN 90-277-0811-8
120. I. Pörn, *Action Theory and Social Science*. Some Formal Models. 1977 ISBN 90-277-0846-0
121. J. Margolis, *Persons and Mind*. The Prospects of Nonreductive Materialism. [Boston Studies in the Philosophy of Science, Vol. LVII] 1977 ISBN 90-277-0854-1; Pb 90-277-0863-0
122. J. Hintikka, I. Niiniluoto, and E. Saarinen (eds.), *Essays on Mathematical and Philosophical Logic*. 1979 ISBN 90-277-0879-7
123. T. A. F. Kuipers, *Studies in Inductive Probability and Rational Expectation*. 1978 ISBN 90-277-0882-7
124. E. Saarinen, R. Hilpinen, I. Niiniluoto and M. P. Hintikka (eds.), *Essays in Honour of Jaakko Hintikka on the Occasion of His 50th Birthday*. 1979 ISBN 90-277-0916-5
125. G. Radnitzky and G. Andersson (eds.), *Progress and Rationality in Science*. [Boston Studies in the Philosophy of Science, Vol. LVIII] 1978 ISBN 90-277-0921-1; Pb 90-277-0922-X
126. P. Mittelstaedt, *Quantum Logic*. 1978 ISBN 90-277-0925-4
127. K. A. Bowen, *Model Theory for Modal Logic*. Kripke Models for Modal Predicate Calculi. 1979 ISBN 90-277-0929-7
128. H. A. Bursen, *Dismantling the Memory Machine*. A Philosophical Investigation of Machine Theories of Memory. 1978 ISBN 90-277-0933-5
129. M. W. Wartofsky, *Models*. Representation and the Scientific Understanding. [Boston Studies in the Philosophy of Science, Vol. XLVIII] 1979 ISBN 90-277-0736-7; Pb 90-277-0947-5
130. D. Ihde, *Technics and Praxis*. A Philosophy of Technology. [Boston Studies in the Philosophy of Science, Vol. XXIV] 1979 ISBN 90-277-0953-X; Pb 90-277-0954-8
131. J. J. Wiatr (ed.), *Polish Essays in the Methodology of the Social Sciences*. [Boston Studies in the Philosophy of Science, Vol. XXIX] 1979 ISBN 90-277-0723-5; Pb 90-277-0956-4
132. W. C. Salmon (ed.), *Hans Reichenbach: Logical Empiricist*. 1979 ISBN 90-277-0958-0
133. P. Bieri, R.-P. Horstmann and L. Krüger (eds.), *Transcendental Arguments in Science*. Essays in Epistemology. 1979 ISBN 90-277-0963-7; Pb 90-277-0964-5
134. M. Marković and G. Petrović (eds.), *Praxis*. Yugoslav Essays in the Philosophy and Methodology of the Social Sciences. [Boston Studies in the Philosophy of Science, Vol. XXXVI] 1979 ISBN 90-277-0727-8; Pb 90-277-0968-8
135. R. Wójcicki, *Topics in the Formal Methodology of Empirical Sciences*. Translated from Polish. 1979 ISBN 90-277-1004-X
136. G. Radnitzky and G. Andersson (eds.), *The Structure and Development of Science*. [Boston Studies in the Philosophy of Science, Vol. LIX] 1979 ISBN 90-277-0994-7; Pb 90-277-0995-5
137. J. C. Webb, *Mechanism, Mentalism and Metamathematics*. An Essay on Finitism. 1980 ISBN 90-277-1046-5
138. D. F. Gustafson and B. L. Tapscott (eds.), *Body, Mind and Method*. Essays in Honor of Virgil C. Aldrich. 1979 ISBN 90-277-1013-9
139. L. Nowak, *The Structure of Idealization*. Towards a Systematic Interpretation of the Marxian Idea of Science. 1980 ISBN 90-277-1014-7

SYNTHESE LIBRARY

140. C. Perelman, *The New Rhetoric and the Humanities*. Essays on Rhetoric and Its Applications. Translated from French and German. With an Introduction by H. Zyskind. 1979
ISBN 90-277-1018-X; Pb 90-277-1019-8
141. W. Rabinowicz, *Universalizability*. A Study in Morals and Metaphysics. 1979
ISBN 90-277-1020-2
142. C. Perelman, *Justice, Law and Argument*. Essays on Moral and Legal Reasoning. Translated from French and German. With an Introduction by H.J. Berman. 1980
ISBN 90-277-1089-9; Pb 90-277-1090-2
143. S. Kanger and S. Öhman (eds.), *Philosophy and Grammar*. Papers on the Occasion of the Quincentennial of Uppsala University. 1981 ISBN 90-277-1091-0
144. T. Pawlowski, *Concept Formation in the Humanities and the Social Sciences*. 1980
ISBN 90-277-1096-1
145. J. Hintikka, D. Gruender and E. Agazzi (eds.), *Theory Change, Ancient Axiomatics and Galileo's Methodology*. Proceedings of the 1978 Pisa Conference on the History and Philosophy of Science, Volume I. 1981 ISBN 90-277-1126-7
146. J. Hintikka, D. Gruender and E. Agazzi (eds.), *Probabilistic Thinking, Thermodynamics, and the Interaction of the History and Philosophy of Science*. Proceedings of the 1978 Pisa Conference on the History and Philosophy of Science, Volume II. 1981 ISBN 90-277-1127-5
147. U. Mönnich (ed.), *Aspects of Philosophical Logic*. Some Logical Forays into Central Notions of Linguistics and Philosophy. 1981 ISBN 90-277-1201-8
148. D. M. Gabbay, *Semantical Investigations in Heyting's Intuitionistic Logic*. 1981
ISBN 90-277-1202-6
149. E. Agazzi (ed.), *Modern Logic – A Survey*. Historical, Philosophical, and Mathematical Aspects of Modern Logic and Its Applications. 1981 ISBN 90-277-1137-2
150. A. F. Parker-Rhodes, *The Theory of Indistinguishables*. A Search for Explanatory Principles below the Level of Physics. 1981 ISBN 90-277-1214-X
151. J. C. Pitt, *Pictures, Images, and Conceptual Change*. An Analysis of Wilfrid Sellars' Philosophy of Science. 1981 ISBN 90-277-1276-X; Pb 90-277-1277-8
152. R. Hilpinen (ed.), *New Studies in Deontic Logic*. Norms, Actions, and the Foundations of Ethics. 1981 ISBN 90-277-1278-6; Pb 90-277-1346-4
153. C. Dilworth, *Scientific Progress*. A Study Concerning the Nature of the Relation between Successive Scientific Theories. 3rd rev. ed., 1994 ISBN 0-7923-2487-0; Pb 0-7923-2488-9
154. D. Woodruff Smith and R. McIntyre, *Husserl and Intentionality*. A Study of Mind, Meaning, and Language. 1982 ISBN 90-277-1392-8; Pb 90-277-1730-3
155. R. J. Nelson, *The Logic of Mind*. 2nd. ed., 1989 ISBN 90-277-2819-4; Pb 90-277-2822-4
156. J. F. A. K. van Benthem, *The Logic of Time*. A Model-Theoretic Investigation into the Varieties of Temporal Ontology, and Temporal Discourse. 1983; 2nd ed., 1991 ISBN 0-7923-1081-0
157. R. Swinburne (ed.), *Space, Time and Causality*. 1983 ISBN 90-277-1437-1
158. E. T. Jaynes, *Papers on Probability, Statistics and Statistical Physics*. Ed. by R. D. Rozenkrantz. 1983 ISBN 90-277-1448-7; Pb (1989) 0-7923-0213-3
159. T. Chapman, *Time: A Philosophical Analysis*. 1982 ISBN 90-277-1465-7
160. E. N. Zalta, *Abstract Objects*. An Introduction to Axiomatic Metaphysics. 1983
ISBN 90-277-1474-6
161. S. Harding and M. B. Hintikka (eds.), *Discovering Reality*. Feminist Perspectives on Epistemology, Metaphysics, Methodology, and Philosophy of Science. 1983
ISBN 90-277-1496-7; Pb 90-277-1538-6
162. M. A. Stewart (ed.), *Law, Morality and Rights*. 1983 ISBN 90-277-1519-X

SYNTHESE LIBRARY

163. D. Mayr and G. Süssmann (eds.), *Space, Time, and Mechanics.* Basic Structures of a Physical Theory. 1983 ISBN 90-277-1525-4
164. D. Gabbay and F. Guenthner (eds.), *Handbook of Philosophical Logic.* Vol. I: Elements of Classical Logic. 1983 ISBN 90-277-1542-4
165. D. Gabbay and F. Guenthner (eds.), *Handbook of Philosophical Logic.* Vol. II: Extensions of Classical Logic. 1984 ISBN 90-277-1604-8
166. D. Gabbay and F. Guenthner (eds.), *Handbook of Philosophical Logic.* Vol. III: Alternative to Classical Logic. 1986 ISBN 90-277-1605-6
167. D. Gabbay and F. Guenthner (eds.), *Handbook of Philosophical Logic.* Vol. IV: Topics in the Philosophy of Language. 1989 ISBN 90-277-1606-4
168. A. J. I. Jones, *Communication and Meaning.* An Essay in Applied Modal Logic. 1983 ISBN 90-277-1543-2
169. M. Fitting, *Proof Methods for Modal and Intuitionistic Logics.* 1983 ISBN 90-277-1573-4
170. J. Margolis, *Culture and Cultural Entities.* Toward a New Unity of Science. 1984 ISBN 90-277-1574-2
171. R. Tuomela, *A Theory of Social Action.* 1984 ISBN 90-277-1703-6
172. J. J. E. Gracia, E. Rabossi, E. Villanueva and M. Dascal (eds.), *Philosophical Analysis in Latin America.* 1984 ISBN 90-277-1749-4
173. P. Ziff, *Epistemic Analysis.* A Coherence Theory of Knowledge. 1984 ISBN 90-277-1751-7
174. P. Ziff, *Antiaesthetics.* An Appreciation of the Cow with the Subtile Nose. 1984 ISBN 90-277-1773-7
175. W. Balzer, D. A. Pearce, and H.-J. Schmidt (eds.), *Reduction in Science.* Structure, Examples, Philosophical Problems. 1984 ISBN 90-277-1811-3
176. A. Peczenik, L. Lindahl and B. van Roermund (eds.), *Theory of Legal Science.* Proceedings of the Conference on Legal Theory and Philosophy of Science (Lund, Sweden, December 1983). 1984 ISBN 90-277-1834-2
177. I. Niiniluoto, *Is Science Progressive?* 1984 ISBN 90-277-1835-0
178. B. K. Matilal and J. L. Shaw (eds.), *Analytical Philosophy in Comparative Perspective.* Exploratory Essays in Current Theories and Classical Indian Theories of Meaning and Reference. 1985 ISBN 90-277-1870-9
179. P. Kroes, *Time: Its Structure and Role in Physical Theories.* 1985 ISBN 90-277-1894-6
180. J. H. Fetzer, *Sociobiology and Epistemology.* 1985 ISBN 90-277-2005-3; Pb 90-277-2006-1
181. L. Haaparanta and J. Hintikka (eds.), *Frege Synthesized.* Essays on the Philosophical and Foundational Work of Gottlob Frege. 1986 ISBN 90-277-2126-2
182. M. Detlefsen, *Hilbert's Program.* An Essay on Mathematical Instrumentalism. 1986 ISBN 90-277-2151-3
183. J. L. Golden and J. J. Pilotta (eds.), *Practical Reasoning in Human Affairs.* Studies in Honor of Chaim Perelman. 1986 ISBN 90-277-2255-2
184. H. Zandvoort, *Models of Scientific Development and the Case of Nuclear Magnetic Resonance.* 1986 ISBN 90-277-2351-6
185. I. Niiniluoto, *Truthlikeness.* 1987 ISBN 90-277-2354-0
186. W. Balzer, C. U. Moulines and J. D. Sneed, *An Architectonic for Science.* The Structuralist Program. 1987 ISBN 90-277-2403-2
187. D. Pearce, *Roads to Commensurability.* 1987 ISBN 90-277-2414-8
188. L. M. Vaina (ed.), *Matters of Intelligence.* Conceptual Structures in Cognitive Neuroscience. 1987 ISBN 90-277-2460-1

SYNTHESE LIBRARY

189. H. Siegel, *Relativism Refuted*. A Critique of Contemporary Epistemological Relativism. 1987
     ISBN 90-277-2469-5
190. W. Callebaut and R. Pinxten, *Evolutionary Epistemology*. A Multiparadigm Program, with a Complete Evolutionary Epistemology Bibliograph. 1987   ISBN 90-277-2582-9
191. J. Kmita, *Problems in Historical Epistemology*. 1988   ISBN 90-277-2199-8
192. J. H. Fetzer (ed.), *Probability and Causality*. Essays in Honor of Wesley C. Salmon, with an Annotated Bibliography. 1988   ISBN 90-277-2607-8; Pb 1-5560-8052-2
193. A. Donovan, L. Laudan and R. Laudan (eds.), *Scrutinizing Science*. Empirical Studies of Scientific Change. 1988   ISBN 90-277-2608-6
194. H.R. Otto and J.A. Tuedio (eds.), *Perspectives on Mind*. 1988   ISBN 90-277-2640-X
195. D. Batens and J.P. van Bendegem (eds.), *Theory and Experiment*. Recent Insights and New Perspectives on Their Relation. 1988   ISBN 90-277-2645-0
196. J. Österberg, *Self and Others*. A Study of Ethical Egoism. 1988   ISBN 90-277-2648-5
197. D.H. Helman (ed.), *Analogical Reasoning*. Perspectives of Artificial Intelligence, Cognitive Science, and Philosophy. 1988   ISBN 90-277-2711-2
198. J. Woleński, *Logic and Philosophy in the Lvov-Warsaw School*. 1989   ISBN 90-277-2749-X
199. R. Wójcicki, *Theory of Logical Calculi*. Basic Theory of Consequence Operations. 1988
     ISBN 90-277-2785-6
200. J. Hintikka and M.B. Hintikka, *The Logic of Epistemology and the Epistemology of Logic*. Selected Essays. 1989   ISBN 0-7923-0040-8; Pb 0-7923-0041-6
201. E. Agazzi (ed.), *Probability in the Sciences*. 1988   ISBN 90-277-2808-9
202. M. Meyer (ed.), *From Metaphysics to Rhetoric*. 1989   ISBN 90-277-2814-3
203. R.L. Tieszen, *Mathematical Intuition*. Phenomenology and Mathematical Knowledge. 1989
     ISBN 0-7923-0131-5
204. A. Melnick, *Space, Time, and Thought in Kant*. 1989   ISBN 0-7923-0135-8
205. D.W. Smith, *The Circle of Acquaintance*. Perception, Consciousness, and Empathy. 1989
     ISBN 0-7923-0252-4
206. M.H. Salmon (ed.), *The Philosophy of Logical Mechanism*. Essays in Honor of Arthur W. Burks. With his Responses, and with a Bibliography of Burk's Work. 1990
     ISBN 0-7923-0325-3
207. M. Kusch, *Language as Calculus vs. Language as Universal Medium*. A Study in Husserl, Heidegger, and Gadamer. 1989   ISBN 0-7923-0333-4
208. T.C. Meyering, *Historical Roots of Cognitive Science*. The Rise of a Cognitive Theory of Perception from Antiquity to the Nineteenth Century. 1989   ISBN 0-7923-0349-0
209. P. Kosso, *Observability and Observation in Physical Science*. 1989   ISBN 0-7923-0389-X
210. J. Kmita, *Essays on the Theory of Scientific Cognition*. 1990   ISBN 0-7923-0441-1
211. W. Sieg (ed.), *Acting and Reflecting*. The Interdisciplinary Turn in Philosophy. 1990
     ISBN 0-7923-0512-4
212. J. Karpiński, *Causality in Sociological Research*. 1990   ISBN 0-7923-0546-9
213. H.A. Lewis (ed.), *Peter Geach: Philosophical Encounters*. 1991   ISBN 0-7923-0823-9
214. M. Ter Hark, *Beyond the Inner and the Outer*. Wittgenstein's Philosophy of Psychology. 1990
     ISBN 0-7923-0850-6
215. M. Gosselin, *Nominalism and Contemporary Nominalism*. Ontological and Epistemological Implications of the Work of W.V.O. Quine and of N. Goodman. 1990   ISBN 0-7923-0904-9
216. J.H. Fetzer, D. Shatz and G. Schlesinger (eds.), *Definitions and Definability*. Philosophical Perspectives. 1991   ISBN 0-7923-1046-2
217. E. Agazzi and A. Cordero (eds.), *Philosophy and the Origin and Evolution of the Universe*. 1991   ISBN 0-7923-1322-4

SYNTHESE LIBRARY

218. M. Kusch, *Foucault's Strata and Fields.* An Investigation into Archaeological and Genealogical Science Studies. 1991 ISBN 0-7923-1462-X
219. C.J. Posy, *Kant's Philosophy of Mathematics.* Modern Essays. 1992 ISBN 0-7923-1495-6
220. G. Van de Vijver, *New Perspectives on Cybernetics.* Self-Organization, Autonomy and Connectionism. 1992 ISBN 0-7923-1519-7
221. J.C. Nyíri, *Tradition and Individuality.* Essays. 1992 ISBN 0-7923-1566-9
222. R. Howell, *Kant's Transcendental Deduction.* An Analysis of Main Themes in His Critical Philosophy. 1992 ISBN 0-7923-1571-5
223. A. García de la Sienra, *The Logical Foundations of the Marxian Theory of Value.* 1992 ISBN 0-7923-1778-5
224. D.S. Shwayder, *Statement and Referent.* An Inquiry into the Foundations of Our Conceptual Order. 1992 ISBN 0-7923-1803-X
225. M. Rosen, *Problems of the Hegelian Dialectic.* Dialectic Reconstructed as a Logic of Human Reality. 1993 ISBN 0-7923-2047-6
226. P. Suppes, *Models and Methods in the Philosophy of Science: Selected Essays.* 1993 ISBN 0-7923-2211-8
227. R. M. Dancy (ed.), *Kant and Critique: New Essays in Honor of W. H. Werkmeister.* 1993 ISBN 0-7923-2244-4
228. J. Woleński (ed.), *Philosophical Logic in Poland.* 1993 ISBN 0-7923-2293-2
229. M. De Rijke (ed.), *Diamonds and Defaults.* Studies in Pure and Applied Intensional Logic. 1993 ISBN 0-7923-2342-4
230. B.K. Matilal and A. Chakrabarti (eds.), *Knowing from Words.* Western and Indian Philosophical Analysis of Understanding and Testimony. 1994 ISBN 0-7923-2345-9
231. S.A. Kleiner, *The Logic of Discovery.* A Theory of the Rationality of Scientific Research. 1993 ISBN 0-7923-2371-8
232. R. Festa, *Optimum Inductive Methods.* A Study in Inductive Probability, Bayesian Statistics, and Verisimilitude. 1993 ISBN 0-7923-2460-9
233. P. Humphreys (ed.), *Patrick Suppes: Scientific Philosopher.* Vol. 1: Probability and Probabilistic Causality. 1994 ISBN 0-7923-2552-4
234. P. Humphreys (ed.), *Patrick Suppes: Scientific Philosopher.* Vol. 2: Philosophy of Physics, Theory Structure, and Measurement Theory. 1994 ISBN 0-7923-2553-2
235. P. Humphreys (ed.), *Patrick Suppes: Scientific Philosopher.* Vol. 3: Language, Logic, and Psychology. 1994 ISBN 0-7923-2862-0
  Set ISBN (Vols 233–235) 0-7923-2554-0
236. D. Prawitz and D. Westerståhl (eds.), *Logic and Philosophy of Science in Uppsala.* Papers from the 9th International Congress of Logic, Methodology, and Philosophy of Science. 1994 ISBN 0-7923-2702-0
237. L. Haaparanta (ed.), *Mind, Meaning and Mathematics.* Essays on the Philosophical Views of Husserl and Frege. 1994 ISBN 0-7923-2703-9
238. J. Hintikka (ed.), *Aspects of Metaphor.* 1994 ISBN 0-7923-2786-1
239. B. McGuinness and G. Oliveri (eds.), *The Philosophy of Michael Dummett.* With Replies from Michael Dummett. 1994 ISBN 0-7923-2804-3
240. D. Jamieson (ed.), *Language, Mind, and Art.* Essays in Appreciation and Analysis, In Honor of Paul Ziff. 1994 ISBN 0-7923-2810-8
241. G. Preyer, F. Siebelt and A. Ulfig (eds.), *Language, Mind and Epistemology.* On Donald Davidson's Philosophy. 1994 ISBN 0-7923-2811-6
242. P. Ehrlich (ed.), *Real Numbers, Generalizations of the Reals, and Theories of Continua.* 1994 ISBN 0-7923-2689-X

SYNTHESE LIBRARY

243. G. Debrock and M. Hulswit (eds.), *Living Doubt*. Essays concerning the epistemology of Charles Sanders Peirce. 1994 ISBN 0-7923-2898-1
244. J. Srzednicki, *To Know or Not to Know*. Beyond Realism and Anti-Realism. 1994
    ISBN 0-7923-2909-0
245. R. Egidi (ed.), *Wittgenstein: Mind and Language*. 1995 ISBN 0-7923-3171-0
246. A. Hyslop, *Other Minds*. 1995 ISBN 0-7923-3245-8
247. L. Pólos and M. Masuch (eds.), *Applied Logic: How, What and Why*. Logical Approaches to Natural Language. 1995 ISBN 0-7923-3432-9
248. M. Krynicki, M. Mostowski and L.M. Szczerba (eds.), *Quantifiers: Logics, Models and Computation*. Volume One: Surveys. 1995 ISBN 0-7923-3448-5
249. M. Krynicki, M. Mostowski and L.M. Szczerba (eds.), *Quantifiers: Logics, Models and Computation*. Volume Two: Contributions. 1995 ISBN 0-7923-3449-3
    Set ISBN (Vols 248 + 249) 0-7923-3450-7
250. R.A. Watson, *Representational Ideas from Plato to Patricia Churchland*. 1995
    ISBN 0-7923-3453-1
251. J. Hintikka (ed.), *From Dedekind to Gödel*. Essays on the Development of the Foundations of Mathematics. 1995 ISBN 0-7923-3484-1
252. A. Wiśniewski, *The Posing of Questions*. Logical Foundations of Erotetic Inferences. 1995
    ISBN 0-7923-3637-2
253. J. Peregrin, *Doing Worlds with Words*. Formal Semantics without Formal Metaphysics. 1995
    ISBN 0-7923-3742-5
254. I.A. Kieseppä, *Truthlikeness for Multidimensional, Quantitative Cognitive Problems*. 1996
    ISBN 0-7923-4005-1
255. P. Hugly and C. Sayward: *Intensionality and Truth*. An Essay on the Philosophy of A.N. Prior. 1996 ISBN 0-7923-4119-8
256. L. Hankinson Nelson and J. Nelson (eds.): *Feminism, Science, and the Philosophy of Science*. 1997 ISBN 0-7923-4162-7
257. P.I. Bystrov and V.N. Sadovsky (eds.): *Philosophical Logic and Logical Philosophy*. Essays in Honour of Vladimir A. Smirnov. 1996 ISBN 0-7923-4270-4
258. Å.E. Andersson and N-E. Sahlin (eds.): *The Complexity of Creativity*. 1996
    ISBN 0-7923-4346-8
259. M.L. Dalla Chiara, K. Doets, D. Mundici and J. van Benthem (eds.): *Logic and Scientific Methods*. Volume One of the Tenth International Congress of Logic, Methodology and Philosophy of Science, Florence, August 1995. 1997 ISBN 0-7923-4383-2
260. M.L. Dalla Chiara, K. Doets, D. Mundici and J. van Benthem (eds.): *Structures and Norms in Science*. Volume Two of the Tenth International Congress of Logic, Methodology and Philosophy of Science, Florence, August 1995. 1997 ISBN 0-7923-4384-0
    Set ISBN (Vols 259 + 260) 0-7923-4385-9
261. A. Chakrabarti: *Denying Existence*. The Logic, Epistemology and Pragmatics of Negative Existentials and Fictional Discourse. 1997 ISBN 0-7923-4388-3
262. A. Biletzki: *Talking Wolves*. Thomas Hobbes on the Language of Politics and the Politics of Language. 1997 ISBN 0-7923-4425-1
263. D. Nute (ed.): *Defeasible Deontic Logic*. 1997 ISBN 0-7923-4630-0
264. U. Meixner: *Axiomatic Formal Ontology*. 1997 ISBN 0-7923-4747-X
265. I. Brinck: *The Indexical 'I'*. The First Person in Thought and Language. 1997
    ISBN 0-7923-4741-2
266. G. Hölmström-Hintikka and R. Tuomela (eds.): *Contemporary Action Theory*. Volume 1: Individual Action. 1997 ISBN 0-7923-4753-6; Set: 0-7923-4754-4

SYNTHESE LIBRARY

267. G. Hölmström-Hintikka and R. Tuomela (eds.): *Contemporary Action Theory.* Volume 2: Social Action. 1997 ISBN 0-7923-4752-8; Set: 0-7923-4754-4
268. B.-C. Park: *Phenomenological Aspects of Wittgenstein's Philosophy.* 1998
ISBN 0-7923-4813-3
269. J. Paśniczek: *The Logic of Intentional Objects.* A Meinongian Version of Classical Logic. 1998
Hb ISBN 0-7923-4880-X; Pb ISBN 0-7923-5578-4
270. P.W. Humphreys and J.H. Fetzer (eds.): *The New Theory of Reference.* Kripke, Marcus, and Its Origins. 1998 ISBN 0-7923-4898-2
271. K. Szaniawski, A. Chmielewski and J. Woleński (eds.): *On Science, Inference, Information and Decision Making.* Selected Essays in the Philosophy of Science. 1998
ISBN 0-7923-4922-9
272. G.H. von Wright: *In the Shadow of Descartes.* Essays in the Philosophy of Mind. 1998
ISBN 0-7923-4992-X
273. K. Kijania-Placek and J. Woleński (eds.): *The Lvov–Warsaw School and Contemporary Philosophy.* 1998 ISBN 0-7923-5105-3
274. D. Dedrick: *Naming the Rainbow.* Colour Language, Colour Science, and Culture. 1998
ISBN 0-7923-5239-4
275. L. Albertazzi (ed.): *Shapes of Forms.* From Gestalt Psychology and Phenomenology to Ontology and Mathematics. 1999 ISBN 0-7923-5246-7
276. P. Fletcher: *Truth, Proof and Infinity.* A Theory of Constructions and Constructive Reasoning. 1998 ISBN 0-7923-5262-9
277. M. Fitting and R.L. Mendelsohn (eds.): *First-Order Modal Logic.* 1998
Hb ISBN 0-7923-5334-X; Pb ISBN 0-7923-5335-8
278. J.N. Mohanty: *Logic, Truth and the Modalities from a Phenomenological Perspective.* 1999
ISBN 0-7923-5550-4
279. T. Placek: *Mathematical Intiutionism and Intersubjectivity.* A Critical Exposition of Arguments for Intuitionism. 1999 ISBN 0-7923-5630-6
280. A. Cantini, E. Casari and P. Minari (eds.): *Logic and Foundations of Mathematics.* 1999
ISBN 0-7923-5659-4   set ISBN 0-7923-5867-8
281. M.L. Dalla Chiara, R. Giuntini and F. Laudisa (eds.): *Language, Quantum, Music.* 1999
ISBN 0-7923-5727-2; set ISBN 0-7923-5867-8
282. R. Egidi (ed.): *In Search of a New Humanism.* The Philosophy of Georg Hendrik von Wright. 1999 ISBN 0-7923-5810-4
283. F. Vollmer: *Agent Causality.* 1999 ISBN 0-7923-5848-1
284. J. Peregrin (ed.): *Truth and Its Nature (if Any).* 1999 ISBN 0-7923-5865-1
285. M. De Caro (ed.): *Interpretations and Causes.* New Perspectives on Donald Davidson's Philosophy. 1999 ISBN 0-7923-5869-4
286. R. Murawski: *Recursive Functions and Metamathematics.* Problems of Completeness and Decidability, Gödel's Theorems. 1999 ISBN 0-7923-5904-6
287. T.A.F. Kuipers: *From Instrumentalism to Constructive Realism.* On Some Relations between Confirmation, Empirical Progress, and Truth Approximation. 2000 ISBN 0-7923-6086-9
288. G. Holmström-Hintikka (ed.): *Medieval Philosophy and Modern Times.* 2000
ISBN 0-7923-6102-4
289. E. Grosholz and H. Breger (eds.): *The Growth of Mathematical Knowledge.* 2000
ISBN 0-7923-6151-2

SYNTHESE LIBRARY

290. G. Sommaruga: *History and Philosophy of Constructive Type Theory.* 2000
ISBN 0-7923-6180-6
291. J. Gasser (ed.): *A Boole Anthology.* Recent and Classical Studies in the Logic of George Boole. 2000
ISBN 0-7923-6380-9
292. V.F. Hendricks, S.A. Pedersen and K.F. Jørgensen (eds.): *Proof Theory.* History and Philosophical Significance. 2000
ISBN 0-7923-6544-5
293. W.L. Craig: *The Tensed Theory of Time.* A Critical Examination. 2000 ISBN 0-7923-6634-4
294. W.L. Craig: *The Tenseless Theory of Time.* A Critical Examination. 2000
ISBN 0-7923-6635-2
295. L. Albertazzi (ed.): *The Dawn of Cognitive Science.* Early European Contributors. 2001
ISBN 0-7923-6799-5
296. G. Forrai: *Reference, Truth and Conceptual Schemes.* A Defense of Internal Realism. 2001
ISBN 0-7923-6885-1
297. V.F. Hendricks, S.A. Pedersen and K.F. Jørgensen (eds.): *Probability Theory.* Philosophy, Recent History and Relations to Science. 2001
ISBN 0-7923-6952-1
298. M. Esfeld: *Holism in Philosophy of Mind and Philosophy of Physics.* 2001
ISBN 0-7923-7003-1
299. E.C. Steinhart: *The Logic of Metaphor.* Analogous Parts of Possible Worlds. 2001
ISBN 0-7923-7004-X
300. To be published.
301. T.A.F. Kuipers: *Structures in Science Heuristic Patterns Based on Cognitive Structures.* An Advanced Textbook in Neo-Classical Philosophy of Science. 2001   ISBN 0-7923-7117-8
302. G. Hon and S.S. Rakover (eds.): *Explanation.* Theoretical Approaches and Applications. 2001
ISBN 1-4020-0017-0
303. G. Holmström-Hintikka, S. Lindström and R. Sliwinski (eds.): *Collected Papers of Stig Kanger with Essays on his Life and Work.* Vol. I. 2001
ISBN 1-4020-0021-9; Pb ISBN 1-4020-0022-7
304. G. Holmström-Hintikka, S. Lindström and R. Sliwinski (eds.): *Collected Papers of Stig Kanger with Essays on his Life and Work.* Vol. II. 2001
ISBN 1-4020-0111-8; Pb ISBN 1-4020-0112-6
305. C.A. Anderson and M. Zelëny (eds.): *Logic, Meaning and Computation.* Essays in Memory of Alonzo Church. 2001
ISBN 1-4020-0141-X
306. P. Schuster, U. Berger and H. Osswald (eds.): *Reuniting the Antipodes – Constructive and Nonstandard Views of the Continuum.* 2001
ISBN 1-4020-0152-5
307. S.D. Zwart: *Refined Verisimilitude.* 2001
ISBN 1-4020-0268-8
308. A.-S. Maurin: *IF Tropes.* 2002
ISBN 1-4020-0656-X
309. H. Eilstein (ed.): *A Collection of Polish Works on Philosophical Problems of Time and Spacetime.* 2002
ISBN 1-4020-0670-5
310. Y. Gauthier: *Internal Logic.* Foundations of Mathematics from Kronecker to Hilbert. 2002
ISBN 1-4020-0689-6
311. E. Ruttkamp: *A Model-Theoretic Realist Interpretation of Science.* 2002
ISBN 1-4020-0729-9

Previous volumes are still available.

KLUWER ACADEMIC PUBLISHERS – DORDRECHT / BOSTON / LONDON